Production Sound Mixing

The Cinetech Guides to the Film Crafts

Series Editor: David Landau

Production Sound Mixing

THE ART AND CRAFT OF SOUND RECORDING FOR THE MOVING IMAGE

JOHN J. MURPHY

Bloomsbury Academic
An imprint of Bloomsbury Publishing Inc

B L O O   S B   R Y
NEW YORK · LONDON · OXFORD · NEW DELHI · SYDNEY

Bloomsbury Academic
An imprint of Bloomsbury Publishing Inc

1385 Broadway   50 Bedford Square
New York   London
NY 10018   WC1B 3DP
USA   UK

www.bloomsbury.com

BLOOMSBURY and the Diana logo are trademarks of Bloomsbury Publishing Plc

First published 2016

Library of Congress Cataloging-in-Publication Data
Murphy, John J., author.
Production sound mixing : the art and craft of sound recording for the moving image / John J. Murphy.
pages cm. — (The cinetech guides to the film crafts)
Includes index.
ISBN 978-1-5013-0709-6 (hardback : alk. paper) — ISBN 978-1-5013-0708-9 (pbk. : alk. paper)
1. Sound in motion pictures.   2. Sound—Recording and reproduction.   I. Title.
TR897.M84 2015
777'.53—dc23
2015025735

ISBN: HB: 9781501307096
PB: 9781501307089
ePub: 9781501307102
ePDF: 9781501307126

Series: The CineTech Guides to the Film Crafts

Typeset by RefineCatch Limited, Bungay, Suffolk
Printed and bound in the United States of America

CONTENTS

PREFACE

When I was first approached by my editor David Landau to write a book about location sound recording, I immediately accepted. After all, I was aware that very few volumes have been written about location sound. It is no surprise, as the vocation of audio has many different types of work one can choose as a sound person. It's like we are talking about chefs. From the backyard picnic burger-flipper (the personal documentarian) to the five-star chef (feature film mixer) they all deal with the same medium and techniques, and face very similar challenges under wildly different conditions and pressures.

This volume is intended for those in the very early stages of discovery of the great process of capturing sound for the moving image. With this reader in mind, I hope to provide an overview, often briefly introducing concepts to which entire works could be devoted. The guiding principle throughout is to shed light on the practical applications of theory rather than the theory itself.

The wide-ranging scope of types of sound work covered in this volume mirrors my own thirty year career. A career that seems to be all over the map, sometime literally.

The level of information presented is calibrated by my experience of teaching 13 years in the film department of an art school. Teaching (and writing) to this level of young people has had the unexpected benefit not only of ordering the thought process but also of asking oneself: "*Why* is it done that way?"

Part of the narrative of this book is the liberal use of my own personal experiences. These are only intended, by example, to carry on in the great oral tradition of pilots and sailors everywhere saying: "I learned from this . . . You can too."

To this day, I still learn something new on every job I work on.

ACKNOWLEDGMENTS

David Landau, my first editor.

Jerry Bruck/Posthorn Recordings, for decades a guiding light of audio knowledge.

John McCormick who, in the beginning, was the one I could call in a panic.

Kira Smith, a stunning example of professionalism.

Alex Noyes, a double threat: an academic who works in the industry.

RESOURCES

A selection of Bresson quotes throughout are from *Notes on the Cinematographer*, trans. Jonathan Griffin (Los Angeles: Green Integer, 1997).

INTRODUCTION

No matter how exotic a camera is, be it film or video, the sound that goes with this stylish image must always be as clear and as natural as possible.

Visual illusions are what make the moving image great. Fluid movement, narrow depth of field, and high speed exposures all constantly transform the images that we watch.

However, if the exotic moving images are of people, and those people are communicating, it really is only a lovely kaleidoscope unless the voices can be understood.

The baseline of any film is the ability of the viewer to understand the dialog of the people on screen.

This book will give the reader the knowledge, in virtually any situation, to record the voices in frame cleanly and naturally.

We who do our work on location deal with many challenges.

We must face the reality that it is not our choice where we work, but a group decision that encompasses the vision of the director and his script, the available locations to fulfill this vision, and the schedule that allows the production to shoot at this location in a given timeframe.

If you are working on a documentary, the location is often the site where a historically significant event took place. It may be a nightmare for audio, but it just has to be there.

For example, you may be working on an industrial film, and a particular machine must be featured in the shot, and the designer must talk about it to camera. In the factory, while the factory is in operation.

Or if you are covering breaking news, this is often noisy, or happening in really loud and difficult audio situations.

There may be hurricanes, riots, and wars, but we still must hear the correspondent ask the questions, and hear the answers.

A location sound recordist has to deal with what most of us automatically correct. In life, in a lively sonic environment, on board a plane, in a taxi, or at a sporting event, we as humans listen and comprehend using intellect to dig out the spoken word from the background gumbo of crazy sound.

We sound recordists cannot depend on this human filter and must get just the intended sound, whether it be words, music, or effects, as cleanly and as naturally as possible.

This book will give those unaware of the sonic world a basic understanding of the physics of sound, and the best ways to capture it.

It will guide the single filmmaker through the minefield of having to do absolutely everything and still record great tracks.

The book will present a number of ideas to those who are just entering the world of professional audio.

Sound has many rewards, technical mastery, logistic wizardry, personal growth, and the unexpected pleasure of finding yourself in situations you would never, ever imagine yourself in. The headwaters of the Amazon? Next week? Sure.

Come and be part of it.

1 WHAT IS SOUND?

"The soundtrack invented silence."

Robert Bresson

Discovering the invisible forces of sound: how does sound work and how do we hear?

Before we can capture sound waves for our moving images, we must first understand its properties and how they affect us. Welcome to an invisible world of tremendous power and boundless possibilities. Sound will become your most powerful tool.

In order to capture the waves, we first must understand them.

WHAT WE WILL LEARN IN THIS CHAPTER

What is sound?

How do we measure something we can't see?

How do we hear?

How does sound energy respond to atmosphere and distance?

Hearing is one of our most vital senses. It helps us communicate, judge distance, and generally connect with our world.

It is also one of our main ways of staying safe, yet it is invisible.

Before we learn to capture sound waves, we must first understand how they work.

Let's go down into the subway.

The subway is a fascinating place, full of interesting science, especially physics.

Never thought of it that way?

Well, sound itself is full of science and physics.

To understand the medium you are dealing with, you have to know a little about how it works.

The subway is full of reverberation.

What's reverberation?

Our Earth is surrounded by an atmosphere of dense gases. When something moves, it moves the atmosphere as well. This movement creates waves. These waves, when processed by our sense of hearing, are interpreted by our brains as sound.

Reverberation is the waves you hear bouncing around after the musician on the platform hits his last note.

The waves echo or linger in the big hollow tube of concrete. It makes his excellent playing sound full and terrific.

It is a product of the space and atmosphere that surround him.

Clap your hands once loudly on the platform. Hear the echo? This is a "live" space.

You have just used the most important tools you possess.

1. Your ears (the organ of hearing).

2. Your own judgement, as to the nature of the sound you hear.

The first you are born with.

The second you develop by being aware of what surrounds you all the time, day in, day out.

We say that this subway tunnel is a "live" space. Interior spaces like this tunnel, studios, or rooms of a house, are categorized by audio engineers as "live" or "dead."

Next, in a place that could not be more different than this subway—back home in your childhood bedroom for example, which your parents have turned into a guest room, with carpet, drapes, and a big quilted bed. Now clap your hands.

Dead. The reverberation after the clap is absorbed by all the soft materials dampening the secondary movement of the sound waves.

Step into the bathroom. Clap your hands. The sound bounces back and forth off the tile, mirrors, and glass.

That's a "live" room.

A little bit of reverberation, or reverb, for short, causes sound to linger. It can reinforce words and musical notes.

That's why our man has chosen the subway to play his music in.

He sounds great! He is playing a steel pan.

What's that?

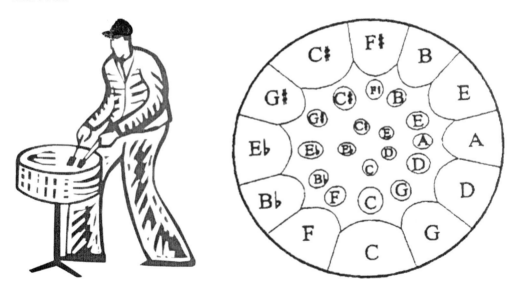

This is a percussion instrument created using the part of a 55-gallon oil drum whose top has been stretched into a bowl shape. Using rubber-tipped sticks, the drummer hits different areas of the bowl, producing a mellow reverberant sound. It is a nice surprise to hear the sound of the islands Trinidad and Tobago deep beneath a city street.

The sound waves of the music he plays are enhanced by the structure of the subway.

It has many hard surfaces and is enclosed as well. The sounds he produces retain a part of their energy by bouncing around from wall to wall.

He is an excellent musician, but playing in the subway makes his music sound fuller and richer due to the reverberation of the concrete walls.

How does this work?

Our drummer hits the surface of the drum with his mallet. The metal *moves*.

As the drum's surface *moves* in the air, it produces *waves* in the atmosphere.

The waves created by the movement of the vibrating metal consist of *alternate regions of high and low pressure*.

These are known as *compression* and *rarefaction*.

Together, these two mirroring motions make up a sound wave.

The sound wave travels through the atmosphere in the tunnel at a high rate of speed.

In dry air, at 20 Celsius (68 F), the sound after he hits his drum travels to your ear at 343.2 meters per second or 1234 kilometers per hour (767 mph).

That's fast!

The primary beat of his mallet on steel speeds to your ear first.
That's *direct sound*. You hear it as one very full note.

Then the radiating beat hits the tile wall and bounces back to your ear a moment later.

This is a *discrete reflection* of the first primary beat. It enhances the first beat by adding a sense of distance or space to the first beat.

It makes it sound "fuller."

The third sound field created is an acoustic mash-up of the first two.

When sound waves are set in motion in a reverberant/reflective space, a full blend of sound waves create an acoustic chorus of primary and decaying waves, each bouncing, sometimes colliding, and sometimes canceling each other out.

This is a full-blown reverberation field.

Reverb, a loose term for sound produced after the fact (note), can be produced either acoustically or electronically. It has been used for years to make singers sound "better" by enriching their voices. That's why you sound so terrific in the shower. *Reverberation*.

MEASURING THE INVISIBLE
We measure waves in the atmosphere by timing them. We clock these waves vibrating over a set period of time.

Heinrich Hertz

That period of time we use is *one second*.

This is the *frequency* of waves occurring over the time of one second, or **how many times** the cycle of compression and refraction occurs (also known as cycles per second) over this period of time.

This system of measurement is named after the scientist who conducted fundamental research in electromagnetism, **Heinrich Hertz**.

We describe sound energy that compresses and refracts 50 times during the course of a single second as having a *frequency* of 50 Hz (50 Hertz).

Sound that vibrates a thousand times a second: 1000 Hz or 1 kHz (one kilohertz).

(The metric measurement term for 1000 is Kilo.)

The steel drum produces sound in a range between the lowest note at 45 Hz and the highest note at 21000 Hz or 21 kHz.

A sound wave's *height or level* is its second defining characteristic. The stronger the vibrations, the greater the *level*, or pressure difference between each compression and rarefaction, and the *louder* the sound.

Most sound **volume**, or **level**, is measured based on the abilities of the human ear.

The range of a sound's volume starts at dead quiet and goes on until your ears reach the threshold of pain.

Volume is measured in *decibels*, designated by the symbol **dB**.

What does this mean? What's decibels?

If we break down the word *decibel* we find:

deci: Latin for 10, and *Bel*, which are the first three letters of another important scientist's name.

Alexander Graham Bell

deci+Bel = **dB** (Decibel) as in Alexander Graham Bell, credited as the inventor of the telephone.

Hence the capital "B" in dB.

Alexander Graham Bell was quite a remarkable individual. His research in sound originated with concern about his mother's increasing deafness. As she began to lose her hearing, Bell would speak in clear, modulated tones directly into her forehead. When even this failed he began to research ways to bring hearing to the deaf. He later taught the deaf, married a deaf woman, designed speedboats, and was a pioneer in early aviation.

Oh, and did I mention he helped start the National Geographic Society? What a guy.

A modulometer measuring sound pressure in dB

THE TWO THINGS THAT DESCRIBE A SOUND

The first is *frequency*: the number of times a sound wave vibrates over one second.

If it vibrates 1200 times in the space of a second we would show it as **1200 Hz**.

The second is *level* or pressure, measured in **dB**. Sometimes this is referred to as volume.

The sound of the steel pan playing a single note again and again in the subway would be described this way:

Steel drum: 1200 Hz @ 85 dB.

What does that mean?

The 1200 Hz is how many times the sound waves vibrate over one second of time—*its pitch or tone.*

Is that high, middle, or low?

If you answered middle, you were correct; 1200 Hz is in the low- to mid-range of what we can hear.

The 85 dB is how loud the sound is. Is that a quiet sound? A normal level of sound? Or is it really loud?

This 85 dB is typical of the loudest sound from a television. It is not only the level of the sound, but the fact he is producing an extremely active sound field. You can hear it over the other noises underground, arriving and departing trains, conversation, the newsagent refilling his booth from a handcart, all due to the reverberation generated by the tunnel walls.

We graphically depict sound in two ways: its *level* and its *pitch* or *frequency*.

We use a chart like the one shown here.

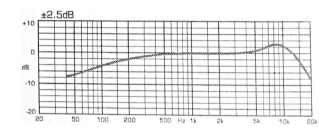

This is a graph that shows how a specific transducer, in this case a microphone, responds to sound pressure.

Notice the vertical line indicates dB.

The horizontal: frequency in Hertz.

How do we hear this?

Hearing is the ability to detect vibrations in the atmosphere through an organ such as your ear.

Your ear detects the vibrations and converts them into nerve impulses that your brain perceives as sound.

The pinna, or visible ear, really is something more than to hang glasses and headphones on. Although the important stuff is inside, the pinna helps to acoustically funnel sound waves into the ear canal, and helps us determine the direction of sounds around us.

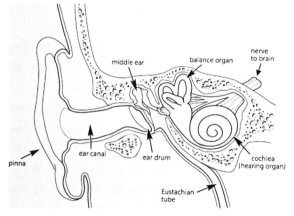

The ear canal is the passageway to the ear drum. The canal is lined with fine hairs and wax glands to keep out objects and dust, and discourage tiny creatures from intruding on the work of the ear drum.

The eardrum is a membrane that is put in motion by sound waves in the atmosphere.

Behind the drum is a fluid filled cavity occupied by three bones that work in unison: the malleus, the incus, and the stapes. When sound waves hit the eardrum (tympanic membrane), a chain reaction occurs moving the three bones, which in turn activates a small oval window at the base of the cochlea, the organ of hearing. Incoming sound waves are magnified by the ear canal resonance, then converted into mechanical vibration by the movement of the eardrum. The lever action of the malleus, incus, and stapes in the middle ear further magnifies the sound. The inner ear converts this motion into electrical signals to the brain via the aural nerves which are connected to thousands of fine hair cells inside the cochlea. It is the motion of these fine hair cells that get converted to nerve impulses that the brain recognizes as sound.

Equal pressure is maintained by a passage called the *Eustachian tube*, which vents back pressure into the nasal passages.

When you feel pressure build up from scuba diving or flight in an airliner, you can clear your ears by yawning, swallowing or chewing gum. All of these actions allow the Eustachian tube to open and equalize pressure around the ear drum.

This passage is named after an Italian scientist who is credited with creating the science of human anatomy.

Within the audible frequency range some frequencies are more invasive than others. The very subway train that rocks our tunnel passes below many buildings on its way uptown. In the dead quiet of night, light sleepers can hear the trains pass beneath their building. The lowest frequencies (30–50 Hz) are the most diffuse frequencies and will pretty much go through anything, like the concrete foundations of buildings. The long powerful wavelengths of the train's passing vibrate the support beams of the buildings along the train's path, and this vibration continues up for many floors to vibrate the atmosphere in that very bedroom.

Bartolomeo Eustachi

High frequencies are much more directional. Additionally we lose the ability to hear high frequencies as we age.

Children and adolescents can hear sounds that adults cannot. Women have a wider range of hearing than men. An eighteen-year-old girl will hear, on average, between 20 Hz and 20000 Hz. A fifty-year-old man will hear a range of sound between 100 Hz and 8000 Hz. Aging causes a progressive loss of hearing, but that can be accelerated by damage to the ears.

Musicians who work in extremely loud environments for extended periods of time will suffer profound hearing loss.

Pete Townsend, leader of The Who, a seminal rock group known for its loud power chords, is just such an example. Townsend, after years of fronting his band, is a victim of hearing loss. He has become a spokesperson for hearing damage awareness. Damage from loud sounds causes the finest of the sound sensing hair cells to break off, limiting your ability to hear. This is very much a function of the time you spend listening to loud sounds. If you listen to 130 dB, the threshold of pain, for one second, it won't harm you. Listen for ten seconds and it will most definitely harm you.

It is extremely important to protect your hearing! Volume must be used in moderation with headphones, particularly the in-ear type.

Once the nerve-ending hairs within the cochlea are damaged, *that's it*. Be careful! After a period of listening to heavy noise everything gradually becomes normal to you. You will turn it up louder progressively without realizing the damage you have done to your ears. Sometimes down in the subway you may notice a commuter in a noisy subway car with their earbuds turned up so loud that they are clearly audible to you sitting far down the car. The terrifying truth is that they won't realize that in four or five years when their hearing begins to fail, the cause can be traced back to this prolonged exposure to loud volumes for extended periods of time. Always use hearing protection around loud concerts, machinery, aircraft, guns, fireworks, or any other source of extreme sound pressure.

A note on guns

It is a fact that in our society and in our visual content, either in an episodic cop show or a documentary, guns play a larger-than-life role. Chances are, sooner or later, you may be called upon to record a gunshot.

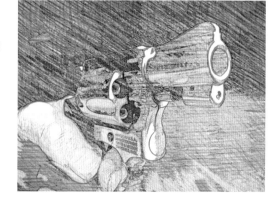

If you point the normal condenser microphone at a discharging weapon, say a handgun, the sound you will record will be most disappointing. It will not sound like a gun at all.

Why not?

The extreme atmospheric pressure of the detonation of the ammunition and its corresponding expulsion of gases will overwhelm your sensitive recording equipment. Both the microphone and the digital recorder will just be overpowered by the onslaught of sonic pressure. The blast of sound will be recorded as a sad pop, or static dropout.

It's just too much.

My techniques for recording gunshots, depending on the environment (indoor-outdoor, near walls, in open spaces), involve pointing the business end of the mike 90 to 180 degrees away from the discharging weapon.

With this technique, you hear the reverberating explosion that sounds just like a gunshot!

Be extremely careful around loaded weapons, even if they are loading blank rounds.

A blank round can kill.

A blank round can deafen a human permanently.

Never allow anyone to point a gun in your direction, loaded or not.

Never take anyone's word that a gun is unloaded.

Treat all guns as if they are loaded and are ready to shoot.

It's a gun. It's a very explicit machine. Its job is to blow holes in things.

Respect the weapon. Stay away from its business end.

Wear hearing protection. Make sure your crew and your colleagues' hearing is protected.

Once you damage your ears, that's it.

It will never heal. Your hearing will never return.

Back to the subway

It's later now and the platform is empty. It's a big empty room with a tunnel at each end.

First of all, with the physics of sound, there is no such thing as an *empty* room. Unless that room is in the *vacuum of space* there will always be atmosphere. That atmosphere is the medium in which sound waves exist.

Without atmosphere, there would be no sound

The subway floods all the time. In fact, it is a constant struggle for the engineers who run the system to keep water out. Water from underground streams, broken water mains, and rainstorms, all try to rush into the space that our trains run through.

Tonight, let's say the tracks have flooded. From the platform you can see a deep river of water sitting a few inches from your feet.

A movement catches your eye, and you notice a rat on a duct above the dark water.

Freaky? *Well, get used to it, down here you are outnumbered.*

The rat runs out of space on the duct, and after surveying her options, dives into the still water between the tracks. Don't worry, she's an excellent swimmer.

Splash. From where she hit, waves radiate outward across the water. That sound, the splash, travels through the air the same way the waves ripple outward from where her whiskers hit the surface.

Each molecule interacts in an orderly way with the other molecules around it.

A given volume of water receives energy from the falling rat and passes it on to other water molecules that are more distant, causing a circular spreading of the wave.

The water responds in a very similar fashion to the way the air is responding to the noise our lady rat/ water surface created.

rattus norvegicus

Her sleek rat-dive disturbed the molecules of the water, but they quickly reoccupied approximately the same positions they had before she took her plunge.

Sound is the same way. The sound she made, the splash, caused a disturbance in the atmosphere of the tunnel. The rat-body plunge caused a *compression* of atmosphere above the water, just as her impact compressed the molecules of water below. That left a vacuum which was quickly filled by molecules of atmosphere. This filling in of the vacuum is called *rarefaction*.

It is similar to the steel drum our man was pounding on this very same platform, only the water is replaced by air molecules around the steel pan. As the metal surface moves away from its still position, it *compresses* the outside air molecules, and as the metal reverses direction to its original position, the air below becomes *rarefied*. This produces a steady tone that the player can rely on each time he hits that particular place on the steel pan.

Look back at the water. Notice the height of the ripples where our rat entered the water. They get *smaller* as they get further from the point she hit, because the energy produced by her impact is being spread over a larger area.

Sound is like this too, *only in three dimensions*. Like someone blowing a bubble, it grows in a 360-degree circle out in all directions, *at the speed of sound*, and *then dissipates over distance*.

The energy of sound waves fall off with distance, just as the ripples get smaller in the water.

The further you get from a source of sound, the quieter it will seem. Its energy is dissipating over distance.

There is a law that describes the energy or amplitude of sound waves reducing with distance.

It's called the *Inverse Square Law*.

When the distance from a sound source doubles, the size of the disturbance is reduced by one quarter.

So the sound of splash, splash, splash of the rat diving contest will be reduced by one quarter if you move away twice as far down the platform from where you stand now.

Let's say you are filming/recording sound at a disaster site. You have to do interviews with the first responders.

There is a generator roaring away to provide emergency lighting.

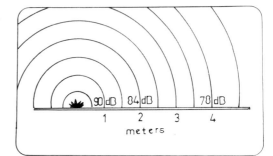

This generator is making your life as a sound recordist really unpleasant. The little engine noise is all over your tracks.

What can I do?
Well, you can't turn off the generator, that's clear. However, you can move away. It seems obvious, but a few meters' distance can make the useless tracks become audible. Also, turn the interviewee with their back between the sound source and your microphone. As the body cast a shadow in the light the generator produces, the body also casts an acoustic shadow, causing the direct sound waves to wrap around the person before getting to the microphone.

Now, as with all laws, there are exceptions . . . Let's say the noise made down here in the subway was not just a lady rat nosediving into the water, but a *hand grenade* going off, and let's say we were off the platform and down the tunnel aways. The explosion would not fall within the inverse square law, because the tunnel walls prevent the sound pressure from spreading. The waves stay concentrated and roll down the tube in both directions. The sound when it gets to you can be nearly as loud as the blast center even far down the tunnel.

Let's go back to the square section of track that has filled with water. When the waves radiated from where our *rattus norvegicus* or lady rat splashed into the water, they moved from her point of impact in an outward direction. Once they hit the concrete sides of the enclosure they bounced back and met the diminishing incoming waves from the original impact. The waves of water, instead of moving in an outward direction merged with each other and began to bob up and down in place.

They became *standing waves*.

Standing waves happen when the sound wave produced fits precisely the dimensions of the space in which the sound occurs.

Like the concrete trough filled with water our rat splashed into, the reflection of the wave off the side reinforces the incoming wave and the pattern seems to stand still, for quite some time, as long as the wave energy is present.

In cathedrals constructed in the twelfth and thirteenth centuries, standing waves produced by organs or choirs had a powerful effect on worshipers standing in the nave (the main worship hall of the church). The dramatic harmonic merging of music into standing waves, coupled with the effects of fasting, must have produced quite a few dramatic moments of devotion among the individual worshipers.

When recording tracks in a room, particularly a small one, standing waves can occur, especially at low frequencies. The room will sound oddly compressed, like you have stuck your head in the very oil drum that our musician's steel pan was constructed from.

How do I get rid of this?
You can move the interview subject within the space, or you can place absorbent materials in the corners of the room where walls meet, or in the place where the floor meets the wall itself. Some well-placed sound absorbing material—blankets, cushions, coats—will break the chain of standing waves.

Doppler shift
You have all heard it. A siren on a passing squad car seems to build as it approaches you, then, having passed, the sound changes or shifts.

In the subway, it is more evident out in the open without the influence of the confining walls of the tunnel. It's called the *Doppler Shift or Effect*.

Doppler did fundamental research in what would happen to wave generators in motion. He postulated that the observed frequency of a wave depended upon the relative speed of the source and the position of the observer. So when the subway train's motorman blows the engine's horn as it pulls into the station, and you are standing on the platform, the sound waves of the horn are crowded together by the speed of the moving train.

As it passes you, the waves spread apart while receding down the tracks. This crowding results in a shorter wavelength as the train approaches, (a higher frequency) and a lowering of the heard frequency as the train recedes.

Christian Johann Doppler

Just going through a phase

OK, we have learned that the two main things that make up a waveform are amplitude and wavelength: Their measurements are noted by the abbreviations dB and Hz. These two measurements are sufficient to describe one sine wave, but they are inadequate to describe all the things in a complex wave?

We need one more concept to describe a wave completely, and that is phase.

To explain phase fully, a number of advanced concepts and properties need to be examined. This discussion is best taken on further in our sonic journey than here at the beginning.

However, how can phase affect you as a sound recordist in the real world?

Noise canceling headsets are quite popular among travelers. However, they should never be used by a sound mixer, for by their very nature of operation they cancel out certain sound waves, and, as you might imagine, a sound mixer is striving to hear all the audio available to him, not less.

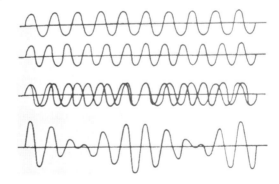

They are also pretty noisy at the higher frequencies.

To help us understand phase, let's examine how they work.

As we have learned, sound is a pressure wave consisting of a compression phase and a rarefaction phase.

A noise cancellation speaker in a headphone earpiece emits a sound wave with the same amplitude but with inverted phase to the original sound. In other words the ambient incoming sound is picked up by a transducer (microphone) in the headphone and digitally processed to be 180 degrees out of phase and directly proportional to the present ambient sound waves.

The two waveforms cancel each other out, effectively reducing the volume of the perceivable noise.

POINTS TO REMEMBER

- The two ways sound is measured are

  - **Frequency, or** *the amount of peaks occurring over one second*

    Long distances between peaks measured over one second means they occurring with a low frequency.

    A short distance between peaks measured over one second, say 5000 waves per second (or 1 kHz: Kilo=1000 +Hz), means a higher frequency of waves occurring represents a higher pitched sound. Like the noise your teapot makes when the water boils.

  - **Level, or** *the height of the sound waves at their peak*

    Sound intensity is called level. This is measured in decibels designated by the letters dB.

    It's not: "Turn up the volume."

    One says: "Increase the level" or "Bring up the level."

- The **inverse square law**: sound energy diminishes over distance.

As the distance from a sound source doubles, the level of the sound from the source is reduced by one quarter.

EXERCISE

Take a blindfold, a handkerchief or one of those sleep aid-eye covers that they hand out on long flights.

Choose five locations of various acoustic live/deadness. Bring a colleague blindfolded into each space and have them clap their hands.

Ask for a detailed description of what they hear:

—*Live? Dead? Reverberant?*

—*What other noises do they hear in the room?*

—*Is this a good place to shoot/record sound?*

2 MICROPHONES:
The Basic Types and How They Work

"A worker may be the hammer's master, but the hammer still prevails. A tool knows exactly how it is meant to be handled, while the user of the tool can have only an approximate idea."

Milan Kundera

There is more to a microphone than being just a sexy object. Each microphone has its own unique capabilities, sensitivities, and personality. Each are designed for specific types of work gathering sound waves. Prepare to learn which microphone, *or transducer*, is right for your audio needs.

Now we will discover a world of unique tools that can capture the waves that fill our world.

Vocalist in front of a Neumann U97 figure of 8 microphone and Close Speech Screen as seen in a camera's viewfinder/monitor

We learned in the last chapter that *there is no sound without movement*.

Hitting an object like a metal drum makes it vibrate. This vibration is transmitted to the air around the object.

This sound causes vibration, and that determines its frequency or pitch.

The frequency at which it vibrates is its first important characteristic.

In the last chapter we also learned how the ear and all its working parts convert sound waves into electrical impulses via nerves to the brain.

A microphone is a kind of ear in that it too converts sound waves into an electrical signal.

It's all about sound waves pushing against a membrane.

In the ear, sound waves push against the eardrum to impart movement to the fluid of the inner ear. This moving fluid causes hair cells to move generating electrical impulses to the brain. These impulses result in the brain's recognition of sound.

In a microphone, the sound waves push a very thin diaphragm or ribbon.

Sound waves move the diaphragm and its motion generates electrical signals.

The amount of voltage produced by the microphone directly depends on the pressure of the sound wave.

All microphones react to the same sound pressure that your ear reacts to.

Electromechanically, microphones are made up of two types: **dynamic** and **condenser**.

Let's look at these two types.

In the dynamic microphone, this reaction moves a coil of wire around a magnet and this generates electrical current which after processing represents the sound which set the diaphragm in motion.

It needs no external power, and its simplicity makes it very inexpensive and durable.

The Condenser microphone has a diaphragm with an electronically charged fixed plate behind it. Electronic circuitry measures the change in distance between the diaphragm and fixed plate and after processing, will represent the sound that set the diaphragm in motion.

A condenser microphone requires power to charge the fixed plate and its corresponding components.

Say that again?

Don't worry, let's go through this step by step. Let's start with . . .

THE DYNAMIC PRESSURE MICROPHONE

This is the simplest and most common type of microphone or *transducer* used in recording sound.

The term *transducer*, although sounding quite scientific, is actually just describing what happens: it converts, *or transforms* energy (acoustical sound waves) into another form of energy (electrical).

Pressure microphones work the same way your ear works.

Stretch a diaphragm over a sealed chamber with a "leak" in the bottom (just like the Eustachian tube in your ear).

A small movable induction coil, positioned in the magnetic field of a permanent magnet, is attached to the diaphragm. When sound waves move the diaphragm, it vibrates, and the coil moves in the magnetic field, producing a current in the coil through electromagnetic induction. This current, although weak, travels via wire to a pre-amp/mixer and then to a media recorder of some type.

This is a *dynamic microphone*. This type is known as a *pressure microphone*.

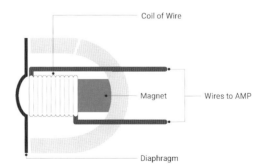

Only one side of the microphone is exposed to the incoming sound waves. The diaphragm responds to the pressure variations uniformly making them inherently omnidirectional.

What does that mean?

Omnidirectional means it picks up sound from all directions.

The simple characteristics of dynamic microphones make them ideal for many situations.

They are inexpensive, robust, and not very sensitive.

Not very sensitive? That's a good thing?

Yes, that's a very good thing in many situations!

What do you mean?

Your local reporter is standing in a field as a storm builds around her. Traffic streams by on a nearby highway. The dynamic mike she uses, say, an Electro Voice RE50, only picks up her voice as she holds it six inches away from her mouth. The building wind, the traffic noise, all have minimal effect on the understandability of her report because really, all you hear is her. The mike she uses is *not very sensitive*.

Perhaps you have been asked to mike musicians for sound reinforcement in a theater or club. If you used ten or so very sensitive mikes for vocals, instruments, and drums, and turned up the loudspeakers, the room would be filled with crazy feedback as the mikes in front of the loudspeaker pick up the ambient sound which will then come out of the speaker, to be picked up by the microphone, and get re-amplified. If the loop gain is sufficient, howling or squealing at the maximum power of the amplifier is possible. This is acoustically amplified *feedback*. *Feedback* is sometimes called *howl-around*, which I think is a more accurate description.

It's universally considered a bad thing when speakers want to be heard. For Jimi Hendrix? Just the thing.

Dynamic mikes, with their lack of sensitivity, can be placed in such an arena and function only when sung or played into due to their lack of sensitivity.

We get an added bonus with the eynamic mike: it is so basic, so simple, it's practically indestructible.

This is why they are embraced by the crews of hard working, hard traveling bands. They are inexpensive, unbreakable, and not too sensitive, just the thing for rock and roll. The dynamic mike can be rattling around in the back of the drummer's janky Econoline van, up and down the Jersey Turnpike, to and from gigs, and it will work just as designed once you plug it in.

In fact, a few years ago ElectroVoice had an advertisement featuring one of their Dynamics actually hammering nails. Yes, nails, and not little finishing nails either—big, build-your-house, ten-penny nails.

The same process that gathers sound in the dynamic mike is used to **reproduce** sound, but on a larger scale.

A speaker is actually all the elements of a dynamic mike but instead of the mike feeding relatively weak signals back to a preamplifier, it is all the same components on a larger scale being driven by a power amplifier. There is now a large composite cone driving the air reproducing sound instead of a tiny mike membrane reacting to the subtle motions of sound waves.

I once attended a baseball game, during which, as well as enjoying the New York Mets on a warm summer night, I had hoped to record some crowd noise. In my bag I had a Sony TCD-3M, a small but high-quality cassette recorder. Toward the end of the event, not wanting to attract undue attention, I connected the headphones and small stereo mike to the TCD-3M. Still enjoying the action and the company of my friends, I placed the headphones around my neck, held up the mike and rolled record on the deck. I looked down and could see the red LED lights moving in sync with the surrounding sound field. All good. I had another adult beverage. Upon returning home, I unpacked my bag and realized that I had somehow switched the mike and headphone plugs.

In other words, I had plugged the headphones into the microphone input and the microphone into the headphone jack. Wait, I saw the LEDs flash . . . I popped out the cassette and inserted it into my stereo system. I rewound and hit play. Out of the speakers came a pretty good stereo recording of what the bleachers sound like at Shea Stadium in the seventh inning.

How could this happen?

The Sony cassette deck provides about 3vDC at the mike input to power its small system of microphones provided for the TCD-3M. This phantom powering brought the headphones to life as a pair of stereo microphones. Right and left headphone speakers became right and left stereo microphones.

Same thing: just different directions.

WHAT IS A CONDENSER?

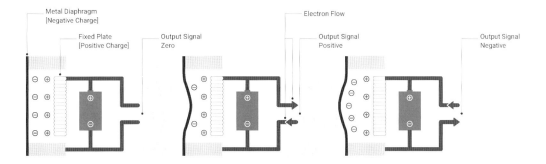

As we have seen, all microphones have a diaphragm that reacts or vibrates as sound waves strike it.

This vibration then causes electrical components to create an output signal.

This microphone has a diaphragm that reacts to the *difference* in sound pressure on either side. It is called a *pressure gradient* microphone.

Unlike a dynamic omni which is a pure pressure microphone, responding to sound from a 360-degree radius around the diaphragm, what the *condenser pressure gradient* microphones all have in common is that their sensitivity depends on the angle to which they are pointed at the intended sound source; *they favor sound that arrives from a particular direction*.

Both sides of the diaphragm are exposed to the sound field, unlike the pressure microphone.

In a ribbon microphone, a thin membrane of foil is suspended between within a magnetic field and registers the difference in pressure between its two sides. This foil diaphragm, because it is reacting to the pressure difference between each side, partially cancels sound waves of equal pressure. It can be useful for this reason in recording dialog placed between two actors in a moving car. A car would have a relatively constant sound field which because it is pounding both sides of the ribbon equally would to some extent, be canceled out. Also the pickup pattern of the figure eight would cover both actors if positioned correctly.

The *condenser* reacts electronically to the incoming sound wave.

It has a front diaphragm, and a fixed back plate.

These plates are charged with a consistent voltage.

Together they form a capacitor.

When sound waves strike the front diaphragm during compression, it moves in, attracting electrons from the diaphragm. Electrons in the output signal flow to the diaphragm. As the diaphragm moves out during rarefaction, electrons in the diaphragm repel each other and flow away from it in the reverse signal.

This signal is sent on its way via a cable to a mike preamp.

Condenser microphones are extremely accurate in the reproduction of sound waves.

They are exquisitely sensitive.

The amplifier of the condenser mike possesses an electronic circuit needed to polarize the capacitor/ diaphragm. This amplifies the signal, making it low impedance and balancing it so it can be sent through a cable. This balancing prevents hum from electromagnetic interference. More about that later.

Let's review

The two types of microphones we have learned about are *dynamic* and *condenser*.

Dynamic mikes are tough, simple, and not all that sensitive. They need no external power because they generate their own signal via electromagnetic induction.

Condenser mikes are sensitive and sophisticated electronic processors requiring a source of power to make their electronics work.

MAPPING THE INVISIBLE

Microphones respond to sound waves in unique ways.

The diagram here is known as a *polar pattern*.

Essentially the polar pattern is a "Map" of how the microphone "picks up" sound at various frequencies.

These patterns are used to graphically depict a microphone's sensitivity as viewed from the side of reception. The microphone's ability to pick up sound waves, or *directionality*, is indicated by the grid overlay indicating front/side/back, with the lines imposed on the grid representing selected frequencies.

This particular polar pattern has a dark line that forms a heart pattern. It represents a cardioid microphone.

The cardioid picks up sound from just the front of the microphone.

The illustrations here are for a figure eight bidirectional microphone.

It is equally sensitive from the front and the back, and not sensitive at all from the sides (or the *null plane*).

The figure eight, with its large diaphragm, is a favorite for vocalists recording in a studio.

Its large diaphragm, and its corresponding sensitivity, make it a standard tool in controlled conditions.

Its usefulness in the field is limited, due to the need to protect it from elements that play havoc with the large ribbon diaphragm. Like protecting it from the wind, for instance. More on wind protection at the end of this chapter.

Polar patterns for other microphone types are as follows.

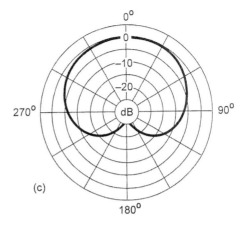

(c)

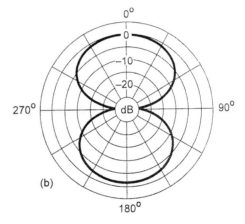

(b)

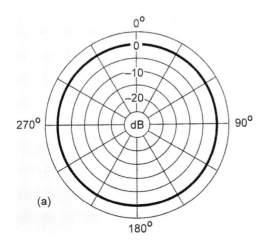

(a)

Omnidirectional: pickup is from all directions around the microphone

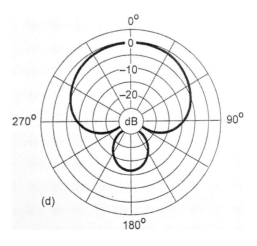

(d)

Hypercardioid

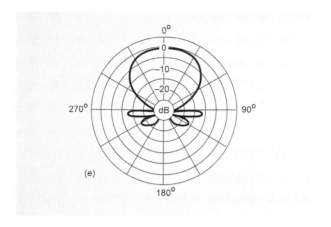

Supercardioid

Condenser microphones are of two types: *Expensive and Really Expensive*.

The less expensive condenser mikes work by having their permanently charged back plate energized by means of a battery.

They are also known as an electret microphone. They require a small amount of power to work, either from a power cell (battery) or phantom powering. They are generally noisier than a true condenser phantom-powered mike, especially toward the end of their useful lives.

The condenser mikes used in film work are powered by a preamplifier in the recording device.

Typically this preamp sends 12–48vDC of power traveling stealthily through the same wires that the audio signal passes through.

This is the reason it is called *phantom power*. The presence of the phantom powering is inaudible in the signal flow. Some versions of popular microphones use T power (which is 12vDC). These were developed for the legendary Nagra analog tape recorder, rarely used today.

Be aware when purchasing used microphones, 12vT equipment may not be compatible with your equipment without a power supply/converter.

The gold standard of microphones lives in the lofty, expensive world of the 48vDC condenser mike.

Names such as Neumann, Sennheiser, and the legendary Schoeps Colette Series are all available for big bucks.

Nagra III

In the world of microphones, you get what you pay for.

If you want the cleanest, most predictable, most reliable performance, spend on the front end.

Your mikes will last a lifetime and never become obsolete.

The top performers are as follows:

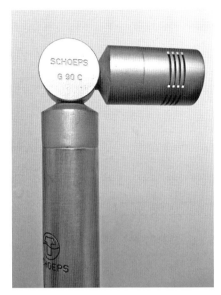

Schoeps Colette System

The Schoeps Colette System mates an amplifier with 20 types of capsules representing any type of microphone you could ever need. A very versatile system.

The venerable Sennheiser MKH 416 P48U3

Sennheiser MKH-60

MICROPHONE TYPES IN USE FOR FILM/VIDEO PRODUCTION

By types we mean not the brand, nor the way it is powered, but the business end of how it functions: *how it picks up sound as shown by the polar pattern*.

Your friend the omni

The most commonly used microphone is the omni or omnidirectional, in the form of either a handheld "stick" mike or the workhorse of interviews and wireless transmission, the *lavaliere*.

In the UK this is called a *"personal mike."*

Virtually all newscasters doing a stand-up in front of camera on the local news, at sporting events, politicians at news conferences, and vocalists in musical productions are using an omni.

It matters not that the mike is in the shot, so the on-camera person holds it right up next to their mouth and delivers the goods.

It's a good thing that the microphone can be seen, for omnis work best when they are nearest the sound source.

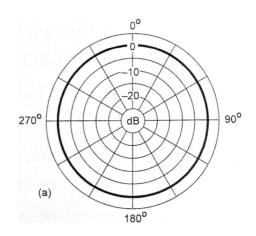

Handheld omni polar pattern

The Electro Voice RE50

To use a very non-audio metaphor, the RE50 is like a fish eye lens.

What is closest to the mike is prominent; the further away you get, the more the sound levels just fall away.

The Electro Voice RE50 is a commonly used, handheld, broadcast microphone.

For dramatic film sound work, using the RE50 on a boom would just not do. You could not get the mike close enough to the talent for meaningful audio levels. However, for a reporter in a crowded stadium, it is just the thing. The mike only picks up the voice, not the voices of the crowd, nor the blur of background music; only what's closest to the mike, the voice, is emphasized. With the addition of a mike flag (the plastic triangle or cube that identifies the station) beneath the head of the mike, the mike on camera is very much a dramatic part of the scene.

The omni lavaliere

From left to right

The Sanken COS-11.

Two Countrymen B-6: the black one is wearing an acoustic cap that promotes "crisp" response.

Two TRAM TR-50s in white and black.

Lavalieres are everywhere. News presenters, talk show hosts, anyone who is up front with a camera on them has a mike clipped to their lapel, tie, neckline, or strap, as they do their work on camera. The mike may be constructed in a color that blends in, but it is there, quite visibly. Sometimes two lavalieres share a tie clip for redundancy in case one fails.

There are two basic types of lavaliere in common usage.

Those that connect to a power supply, then to an audio cable feeding a mixer or camera are known as *hard-wire lavalieres*.

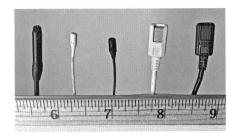

This is because they are *hard-wired* to a connector and then plugged into a camera or mixer/recorder.

The Sennheiser MD 214 from the late 1960s was a sweet-sounding lavaliere that had none of the "chesty" presence of other mikes of this type. The heavy housing also minimized clothing rustle by isolating the transducer deep within the metal frame. It is an example of a dynamic hard-wired lavaliere microphone.

As technology progressed, the lavaliere evolved from dynamic to electret/condenser (see above).

The mike receives power from a battery in a power supply that acts as a receiver for an audio cable or it is powered from the equipment it is connected to.

The second type is identical but has no power supply and has at the end opposite the microphone a miniature audio connector to attach it to a *wireless transmitter*.

(We will cover *Wireless Systems* in Chapter 8.)

In dramatic and some reality television, the microphones must be hidden at all times. This can be problematic to say the least. As I mentioned before about windscreens, no microphone is happy buried under layers of cloth.

They all perform differently under layers of clothing. Some mikes work better than other types when concealed. They must be isolated from fabric contact, human perspiration, the tug and pull of the wearer's activity, and surrounded by a little air for them to do their job.

(We will cover basic mike concealment in Chapter 8.)

The first truly miniature lavaliere was the Sony's ECM-50 (omnidirectional) first produced in 1969.

It quickly became an industry standard.

The ECM-50 was a sweet-sounding mike that was at the beginning of the miniaturization revolution in electronics.

Things keep changing.

Illustrated are three generations of lavaliere: [Top to Bottom] The huge MD 214 from the 1960s, the Sony ECM-50 from the 1970s, and the contemporary Countryman B-6 testify to this revolution continuing today.

Most lavalieres are designed one of two ways: top address or side address.

Sony ECM-50 (1969) and the Countryman B6 (current)

What does that mean?

Look at the Sanken COS 11. It looks like a miniature omnidirectional microphone. The end of the mike picks up the sound.

Now examine the TRAM. Its diaphragm is physically on the flat front of the microphone. It picks up from the *side* of the mike. This can be a great advantage when concealing the microphone under the placket of a shirt, or in a scarf. The design of the plastic structure of the mike protects the sensitive diaphragm from clothing rustle, and allowsyou to be creative with foam and tape to affix it under clothing.

From left to right

Countryman B6

Sanken COS-11

TRAM TR-50

Lectrosonics M152

A good sound person has a selection of lavalieres to use in different situations on various types of clothing.

It's good to have a choice.

The heavy lifters: cardioid/hypercardioid/shotgun

These three condenser microphone patterns are the primary tools used in virtually all forms of film/video production.

At the end of a boom, stashed on a table, or inside a vehicle perched between the sun visors, they are on the receiving end of all quality audio.

The cardioid

Cardioid: the name is derived from the Greek word for "heart." As you can see, the pattern is heart shaped.

As you look at the polar pattern, you can see that cardioid microphones are more sensitive to sound waves in front of the microphone than in back. This directionality is a very valuable tool in productions where the sound source, usually a person, needs to be

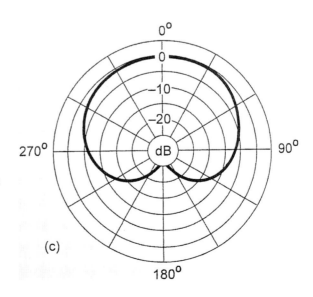

emphasized. The frontal pickup pattern is much less sensitive to noise behind the microphone, such as a noisy camera, the crew, spectators, a squeaky camera dolly, or any other distraction within the sound field.

(We will cover their shooting applications as well as their big brothers, hypercardioid and supercardioid in Chapter 7.)

The hypercardioid

Also known as the supercardioid, this mike is strongly directional. Sound arriving off axis is reduced even more than with a cardioid.

The pickup is "drier" and less susceptible to acoustic reverberations than the cardioid. Its directionality is highly independent of frequency, so that even sounds bouncing off ceilings and walls are registered without coloration. Even when used at a distance from the sound source, it produces a very natural pickup sound. It is an effective space-saving alternative to "shotgun" supercardioid microphones, and because of its small size it can often be placed closer to the sound source.

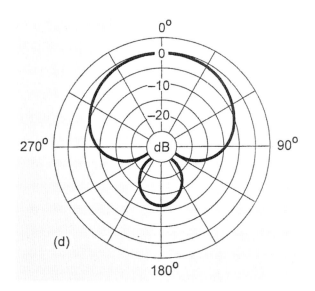

(d)

The supercardioid or shotgun

When recording dialog in active sound fields/noisy spaces, the hypercardioid may pick up too much noise from nearby sources.

This is where the "shotgun" comes into its own. Properly called an *interference tube microphone*, it has small lobes of sensitivity to the left, right, and rear, but is significantly less sensitive to the side and rear than other directional microphones.

This is achieved by placing a slotted tube in front of the diaphragm.

These slots along the tube's sides create wave cancellation, which eliminates much of the off-axis sound. Indoors however, the interference tube mike is notoriously sensitive to its position in a room, where the shifting patterns of acoustic reflections cause corresponding shifts in sound color. Reflectance and reverberations within a

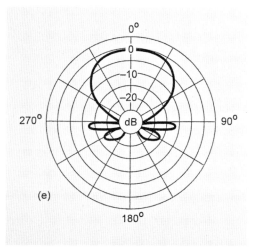

(e)

room contribute greatly to the quality of any sound. A shotgun's pickup angle at high frequencies will be narrower than at lower frequencies. The resulting room sound will be picked up with a distinct high-frequency rolloff, which can make the sound dull or "tubby sounding."

I know only one definitive source to turn to about microphones—Jerry Bruck.

INTERVIEW WITH JERRY BRUCK

Jerry has been recording audio for film and classical music for record release since the 1970s. He is a pioneer of stereo/surround sound, creating the Bruck Surround Microphone System for Schoeps Microphones.

I asked Jerry about the pros and cons of hypercardioid vs. shotgun usage.

What are the differences between a hypercardioid and a shotgun?

One of the reasons we choose a conventional microphone over the shotgun is you don't want to record the whole world, you want to record someone two feet away and that's basically what you are after. There are a lot of good microphones that will do that. A shotgun microphone is actually over the top in that kind of situation, unless you really have a problem with noise in the room, in which case having a microphone that has very directional characteristics is a plus.

I'm not sure the shotgun is always even the answer to that. In fact, you have to put up with the limitations of a shotgun, which often enough are a somewhat strappy sound depending on what microphone we are really talking about. In fact it's a little un-maneuverable as microphones go, because we are talking about one that is twice as long as a normal microphone and less easy to manipulate.

The problem with shotgun microphones is that they are very vulnerable to what's happening around the microphone. What happens with the shotgun microphone is you are picking up enough of the room that it will "color" the sound rather substantially as you move the mike.

You can move it from here to over there and it will give you quite a different sound depending on what reflections it is picking up from the room.

The problem with the shotgun is that it is really not very selective at all about what it's picking up.

It's rare that you have a shotgun that the pickup is somewhat controlled.

The Schoeps CMIT is a good example. You can count on it to represent accurately what things sound like in that room and not be effected by the fact that if you move the microphone it's going to change the sound of what you are picking up.

If you are talking about an inside job, a shotgun is more vulnerable to room acoustics.

Outside is a little different, it's one of the places where shotguns tend to be used when they are not needed.

A lot of times a shotgun is used out of doors, and out of doors is the time when you don't need the directionality.

The tendency for most people is to overdo the quality of pickup. It depends on how your ear is reacting to what you are hearing. Are you really hearing the sound you want to have on the recording, or are you hearing too much of it?

The whole question of selectivity in miking is to try and pick the sound that is representative of the room that you are in and gives you a nice quality that is pleasant to the ear, because it sounds very natural, it's not over the top, and it's not too distant.

That's one of the things that make sound recording an interesting thing to be involved with. You have to constantly ask yourself: "Are you getting the sound that you want, or is it a little too much or not quite enough for what you are trying to get in terms of intelligibility?"

Intelligibility is usually the measure of what we are getting. Is the sound really intelligible?

Are you forcing elements out of that sound that are the ones you want?

You do have to be careful when placing the microphone that you get a natural sound that gives you a feeling of what I would call, ambience, a sense of "Yes we are actually in a room with this person," but not having a problem with other sound that might mislead us or draw us in a different direction. You always look for something that makes the sound seem real. A lot of times I find you get a recording that sounds too present for what it should be.

One of the things you learn about when doing sound recording is to get the right quality of sound.

It's another situation where shotgun microphones tend to overdo things a bit. If you are using a shotgun microphone somewhat carelessly, and you are getting a very close pickup on the voice, it just simply doesn't sound very good. That's not what you are after.

You are looking to get a good voice quality.

So essentially you are looking for the truth, when you are recording sound.

I think that is what you are looking for. You want a quality to the sound that opens up your ear, so to speak, in terms of the naturalness of the recording. The nicest thing to think about when you are making a recording is: Is this the way I really want it to sound in the film, on the recording? Am I overdoing some aspect of it? Or am I not doing enough to the get the quality that I want? Your recording should sound natural.

The only problem with a lot of locations is you may get natural sound all too easily, but some of that natural sound is distracting.

The essence of getting really good vocal quality is achieving something that sounds so natural and so right for the situation. That's all you care about. That's one of the things a sound recordist has to do well, is to get the naturalness of vocal quality.

I hear the room but I'm not distracted by it.

This is another argument for using a boom rather than a lavaliere, because it is about the most natural sound you can get, where using a wire, you have no presence of the room because the mike is on the actor's chest.

Yes, one of the greatest arguments for using the boom is it gives you the most natural pickup you can expect to get. Again, the only problem is, you may be getting more than the sound that you want. That's one reason lavaliere microphones are used in a lot of situations where you are trying so hard to keep out noises that distract you from what you want to hear that the only way to make sure that you have what you want is to over-mike it, getting more of the voice quality on the recording.

When booming, it's very important to track or follow the speaking actor precisely, especially if the mike is close to the speaking actor. Any off-axis sound will be picked up and be noticeably lacking in high-frequency content. In other words, it will sound noticeably flat and hollow if you drift from the sound source/speaking actor.

It takes practice to boom a moving subject (more on this in Chapter 6).

There are two versions of the shotgun: long and short.

Sennheiser pioneered the modern shotgun producing the MKH 416 short shotgun and the MKH 816 long shotgun, both mainstays in the world of film sound.

When you are first starting out as a recordist, you tend to expect greater sensitivity or "selectivity" with a shotgun.

Surprisingly this is not often the case. They cannot isolate one person's voice from another standing a few feet away, or overcome great distances when placed far from the sound source. In fact when placed too far away the sound quality of a shotgun may be distinctly inferior to that of a good hypercardioid, since it will be working in a varied sound field where its response falls off at high frequencies.

Sennheiser MKH 416

It's true that for high-frequency sound, shotguns increase the distance where you can obtain a good proportion of direct sound. Using them really only makes sense when the shotgun is close enough for direct sound to be most prominent.

Schoeps CMIT 5U

PROTECTING THE MICROPHONE FROM WIND

What continually surprises me is what you see at indoor press conferences. There are often four or five booms overhead, each wearing a big fuzzy windscreen.

Inside. Inside a building. Like the White House.

Expecting some heavy weather in the West Wing?

Do you wear earmuffs? Do you take them off when you get indoors? You can hear better, right?

The forces that move the diaphragm in your microphone exist in nature. The same atmospheric waves that are moved by speech are, in nature, gigantic, slow waves that are called wind. This wind must be isolated from the microphone mechanically by the use of barriers—screens, housings, foam, mesh, cloth, fur, and so on—is all part of the technology to isolate your microphone from the wind.

There is no magic wand that makes windscreens acoustically transparent. If you put a big foamy/furry thing on top of your microphone it's going to affect its performance.

Again, put a wool cap over your ears, and it will affect how you hear.

Let's see how windscreens affect a microphone's performance.

Most pop screens or wind screens will have some effect on sound quality.

Close speech screens are for indoor use when a vocalist is singing or speaking closely into a studio microphone.

The construction usually consists of a round frame with nylon mesh stretched over it.

<u>Effect on sound quality:</u> minimal.

<u>Effect on polar response:</u> none.

Popscreens are for indoor use only, to reduce the explosive air currents when speakers pronounce consonants such as "p" or " t."

These are also used when booming indoors.

<u>Effect on sound quality:</u> minimal, restricted to high frequencies and a slight unevenness of response is introduced.

<u>Effect on polar response:</u> minimal and limited to the high frequencies.

Windscreens: (aka *zeppelins*) suppress the disturbances due to relative air motion that you encounter outdoors.

<u>Effect on sound quality:</u> for hypercardioid or shotgun microphones, there is some reduction of low-frequency response due

the pressure difference between the front and back inlets of the capsule which are made smaller by the basket structure. At high frequencies, a slight rolloff and a certain unevenness of response is introduced. This all depends on the microphone type, the diameter of the mesh, the position of the mike in the enclosure, and the possible use of a cloth or fur cover on the zeppelin.

<u>Effect on polar response:</u> at low frequencies the degraded pressure gradient will cause the directional pattern to move toward omnidirectional.

In extreme cases, the supercardioid may take on a cardioid pattern at low frequencies, while a cardioid may widen or become an omni.

Bottom line: all that stuff you put around your mike will affect the sound.

However, if you show up at the scene of the action and the deal is going down, you get the boom up no matter what is on your mike.

I've been that guy. Just get the sound. If you have a second . . . swap out the windscreens.

When inside, take it off. You most likely will never need it. It also makes you look lazy. So unattractive.

POINTS TO REMEMBER

The two basic ways microphones function electromechanically are as follows:

- Dynamic = a simple microphone that responds to sound waves and generates an electrical current. This current reflects the sound waves that moved it and when recorded or amplified gives an accurate reproduction of sound energy.

- Condenser = a microphone that through electronic processing transduces the sound waves that move its diaphragm into an accurate reproduction of the same sound energy.

The basic types of microphones are categorized by the pattern of their sensitivity in relationship to the diaphragm of the microphone itself. This sensitivity is mapped by the use of polar patterns, patterns which show the physical shape of how the microphone responds to song at various frequencies.

The basic fields are:

- omni

- figure of eight

- cardioid

- hypercardioid

- supercardioid.

All non-dynamic microphones require powering to function.

Microphones are constructed with various missions in mind:

- The boom mike for out of frame placement in the course of sound pickup.

- The handheld microphone for close miking the speaker's mouth.

- The lavaliere for chest/body mounting on an individual visibly or hidden.

EXERCISE

Take six microphones and place them on a table.

Identify each and its polar pattern.

Determine if each needs power to operate.

Take each one by one and, with the microphone connected to a mixer, and with your back turned while someone else uses them, identify each microphone by its sound.

3 THE DSLR SHOOT:
Making the Most of What's at Hand

"Someone who can work with the minimum, can work with the most. One who can work with the most, cannot, inevitably, with the minimum."

Robert Bresson, *Notes on the Cinematographer*

Starting out: your first shoot. Let's begin with the basics.

Let's go make that short project you have been dying to shoot!

You have just gotten your first digital camera that can take high definition movies!

My camera does sound, right?

Well, yes, that's true. Most DSLR cameras have a built-in mike.

If you wish to locate it, just look for the tiny holes somewhere on the top front of the camera.

Tiny holes?

Yes. You have just bought a still camera that has the awesome capability to shoot HiDef. That ability was really a bonus option, an add-on. Your camera was conceived and designed for something other than what you are using it for now. It started life as a Stills camera. The HiDef video was really icing on the cake. The DSLR revolution is indeed a disruptive technology in that the ability to shoot HiDef coupled with some pretty awesome lenses suddenly gives you access to a very high level of image production that was only available to professional cameras costing tens of thousands of dollars more.

However, films are not just image making.

Stories are told with another very important component: *sound.*

Behind the little holes on the front of your camera is a tiny omnidirectional mike which, as you remember from the last chapter, best responds to the things nearest to it. That would be you, the cameraperson, and your hand rubbing right over where the mike is as you try and pull focus.

Or the camera strap sliding around, or the lens cap on its little lanyard banging around, or your hands moving to the lens, or the lens itself, buzzing back and forth in autofocus.

I won't need the camera mike because a friend lent me a mike.

Great! What kind of mike? Did they lend you a cable too? What connectors are on the ends of the cable?

Most DSLR cameras have a little plastic tab which underneath is a very familiar hole. It is a 1/8″ Stereo input just like the one in your smartphone or personal listening device.

The 1/8″ stereo jack/plug is probably the most common connector combination in the world. It's also known as the mini plug, or minijack.

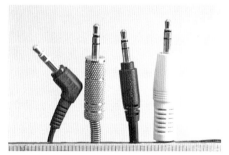

Remember the time you were listening to some tunes and you reached for something and your headphones became unplugged from the little mp3 player you had in your pocket? Silence. The plug slipped right out of that jack. Connection lost.

This is a big disadvantage of our friend the mini plug. It does not lock positively to the jack. There are some mini plugs designed to have locking collars but they require that the host device have a locking ring around the input to work. Few cameras have this type of connector. The examples above are what you encounter in the real world.

What kind of cable did your friend lend you for the microphone?

It's long, but the cable fits the microphone but not the camera.

Professional audio cables have a standardized connector. That connector is the XLR.

What does XLR mean?

When the XLR was originally invented by James Cannon, it was known as the Cannon Plug.

The first series was called Cannon X.

Later designs employed a locking mechanism and were known as XL.

Evolution continued and the female version of the plug had rubber around the holes, XL-R for rubber, so the whole connector became the XLR, now, the standard of the industry.

So, I need an adaptor? Why aren't there cables with an XLR on one end and a mini plug on the other?

An XLR is a *balanced* line. The mini plug is *unbalanced*.

What does that mean?

Conventional consumer electronics wiring is *unbalanced*. This means there is one signal conducting wire inside a shield or braid. Then the whole thing is wrapped in an outer insulator.

This insulator acts as a ground as well as protecting the signal from electrostatic interference.

Microphones, particularly dynamic mikes, as we learned in Chapter 2, operate by generating low voltage levels. A long microphone run with an unbalanced line is really susceptible to electromagnetic interference (hum).

Balanced wiring provides a means to reject this hum caused by stray magnetic fields.

A basic balanced cable contains two identical wires running in parallel with a third cable as ground, these are all wrapped in braid which forms a shield. Two wires form a circuit carrying the audio signal, one in phase to the source signal, the other reversed in polarity. The amplifier measuring the difference in voltage between the two lines rejects any noise that is identical.

The separate shield of a balanced line is better at noise rejection over an unbalanced two conductor cable, such as your mini plug cable where the shield must also act as a signal return wire.

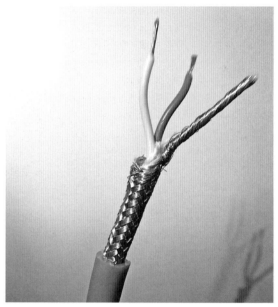

Unbalanced (left) and balanced (right)

The mike my friend lent me is a condenser mike. I plugged it in and it doesn't work. Why?

What is a condenser mike again?

For this mike to operate, it needs power. Some condenser microphones, as we have seen in Chapter 2, are electret microphones and require AA or another type of miniature battery to be inserted into the microphone body.

Other more advanced (and more expensive) microphones require 12–48vDC phantom powering, which must be provided by an external power supply.

So to use that hot condenser mike, you will need a box that provides power: *a power supply.*

How do I control the sound coming in? You know, so it's not too loud?

Most DSLR cameras have an audio recording page in the menu. You are often given a choice of Manual Audio Recording, also known as *Automatic Level Control or ALC.*

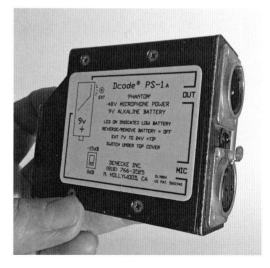

Deneke PS-1 9 volt Power Supply

If you choose the MANUAL setting, you plug in your microphone to the 1/8″ jack of the camera, and set the levels to a safe-midrange place and hope it doesn't get too loud or too quiet.

If you use the automatic level control (ALC) setting on your camera in an environment with a large dynamic range of sounds, for example, a quiet room with normal voice dialog, it might sound pretty "roller coaster." This is when the ALC senses a loud sound and adjusts for it. Then, in the quiet that follows, the ALC pumps up the volume trying to keep the levels constant. It makes for a track full of "whooshes." Not good.

I have an adapter to put the mike mount on the top hot shoe of the camera.

Great! That works if you are right on top of the subject in a face to face sort of situation. If the person is a few meters away, you will not hear them so well, especially if it's in a noisy urban area.

Another setback of the mike on the camera is that it is not really designed to be there. In the heat of shooting, regrettably, the mike may droop, and the fuzzy windscreen that protects the mike from the elements dips into the shot. Grey, out of focus, fuzz in the top of the frame.

I'm not saying you cannot get great results by recording directly into a DSLR. It is just more difficult to do so under a wide range of situations.

What should I do?

Many DSLR videographers use *double system*. This means using a separate recorder for the audio.

Double system refers to the camera/sound recorder combination.

Buying a small multitrack recorder allows you a great deal of versatility at a competitive price. In fact, the recent popularity of DSLR production has fueled, in just a short period of time, a surge in development of new models of this type with many attractive capabilities.

You can control levels, monitor the sound you are recording, power your condenser mikes with 48 volts, record more than one mike, record in stereo (more on this in Chapter 14), and not to be overlooked, record sound, *without the camera.*

The one you purchase should have XLR inputs to allow the use of professional, balanced line, microphones.

Another option is to use a break-out box.

This is a dock for XLR inputs that interface with the camera, sending the inputs via a minijack to the camera's input. They possess some features (powering, bass attenuation, level control), but cannot record audio on their own. Currently, you can purchase a full featured multi-track recorder for the price of a break-out box, so really, the recorder option is much better, for it allows the film maker much more flexibility.

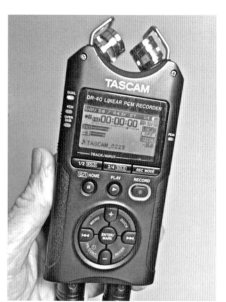

XLR Inputs for external mikes on the bottom of a digital multitrack recorder

OK! Let's shoot!

Find a nice photographic place to put your subject. Make sure it is quiet. Place your subject. Frame them up.

1. Roll sound first.

2. Roll camera.

3. Slate it.

What does "Slate it" mean?

When film first began, a small slate chalkboard was used to identify each shot. It shows the production name, the date, the scene, roll number, and other information important to the production.

When sound arrived in mainstream film production, a noise producing device, *a clap stick*, was added to the slate.

This was a hinged stick on top painted with stripes that is opened and rapidly closed creating an audible noise, a clap.

In the UK the slate is known as a *clapperboard*.

The noise it makes is recorded on the soundtrack.

Later, in post-production, the editor will look for the frame that has the top part of the hinged stick with stripes matching the bottom set of stripes just when they come together. Then the editor finds the sound of the impact of these two pieces of wood as they come together and matches it with the above image. The editing machine is locked and the sound and picture run as one. It has been *synced*.

For your modest DSLR shoot, you are not expected to have a slate; however, a piece of paper with information on it can be very helpful.

What is even more helpful is to have someone clapping their hands together in view of the camera. Or tapping the mike in frame with the camera rolling. You have the visual image in frame to match the audio impact of two hands coming together. This will be an aid in syncing your double system audio with the images recorded by your DSLR. Let's hope your finished project gets more applause than your slates.

If you are doing everything yourself, a *one-man band*, so to speak, it is your responsibility to monitor sound as you shoot. You must do this using headphones.

A word about headphones: Find a pair of headphones that really suit what you do.

What to look for
Headphones that:

- fit over the ear, isolating what you are recording from the outside world;

- have a flat frequency response.

What does this mean?
Look at the beautifully flat frequency response curve depicted on this chart. There is no unnecessary boost at the bottom (bass) end, just a nice flat line showing equal response throughout the operation of this device.

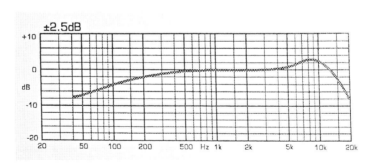

Some headphones are biased for the type of music that the consumer is assumed to be a listener of. Usually there is an emphasis on bass response. In other words, the headphones artificially make the base prominent. Not good. Choose a neutral set of cans (headphones) that give you the most accurate aural reflection of what you are recording. Spend money. These are your main audio reference. Stay with them. Don't change headphones like pairs of shoes. Get to know what sound sounds like through them.

Buy at least two pairs.

Why is this important? Can't I just roll the recorder and forget about it?

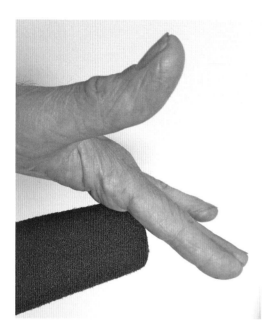

This is a bad idea. Just as things change visually in the frame, your subject might move and you might have to adjust focus or frame, the audio may need attention as well. Your subject is now leaning forward, causing the lavaliere on his lapel to rub against the material of his jacket. Unless you listen, you will never be aware of the fact that your audio has just become kind of unusable.

It's terrible when that happens.

An audience will tolerate a soft or fuzzy shot, but as soon as the voice becomes inaudible, you have lost them.

What can you do to fix this? Subtitles?

That is so sad. When you see a show with English-speaking subjects, and the sound is so bad that the producers have to resort to subtitles.

Remember, in many films, an interview is established with the subject in frame, and an editorial choice is made to cut away to Images of whatever the subject is talking about with the audio track beneath it. You have to have the audio, clear and understandable, for it becomes your film's most important driver.

Why else should you listen?

Well, what else can go wrong? If you are using a wireless lavaliere, your battery may fail, leaving you with a really nice shot, but *silence*.

If outdoors, the wind may increase, causing a terrible rumble as it blows across the microphone. Even if you are working indoors, your interview subject may, in the process of speaking, project a burst of air that causes an annoying pop on the microphone diaphragm.

Always listen!

You have just concluded your interview. **It is very important to check playback.**

Play back the audio to be sure you have sound.

Why?

How else do you know if you have recorded image and sound?

Check it before you leave. You may never have another chance to talk to this person again.

What else should I do?

Record room tone.

What's that?

A really essential part of a sound person's duties.

Room tone is a recording of the atmospheric sound on the location where the interview, scene, shot, stand-up was filmed.

This is typically done on the conclusion of the interview/shooting. It is invaluable to the editorial process. It can cover unwanted noises, as well as help editing voice-over narration once you cutaway from the image of your subject.

Get the room tone.

If your production is strict about take numbers, all sound effects, including room tone, will start with a one thousand series number:

"Effect 1001, 30 seconds room tone, Smith interview, starting now . . . ."

If you are recording on your own double system/standalone recorder a beep of 1 kHz tone at the head and tail will allow the editor to find the room tone audio quickly.

If you are feeding a video camera, at the end of the interview, point the camera at the microphone and roll for 30 seconds. You point the camera at the mike so the editor knows you are doing sound/room tone.

What time is the shoot? This is most important. If an interview starts at 4 p.m. and runs for an hour and a half, you will find in urban environments that rush-hour will start and things will become problematic (noisy). You may want to get room tone first, before shooting, and even afterward.

Is that not asking a lot?

You have to insist on room tone. Often you get little sympathy or tolerance when asking for an entire film set of your fellow craftsmen to be quiet, utterly still, not talk, move, or make any sound. Very often it is the last thing done on set before wrap. That's when it's hardest to restrain those who want to finish and go home. Similarly, if it's an interview and the subject has limited time, or patience, still, get them to be still.

Get that tone.

Ed Bradley, a longtime *60 Minutes* correspondent, usually would tolerate about ten seconds of room tone before he would hold the ticking stopwatch he used for timing video cassettes during interviews right against the lavaliere on his tie. He was telling the sound man he was done.

If they won't wait for your tone, let them leave, then place the mike on the chair at approximately the same level that it was worn. It may not be a scientific match, but it's better than nothing.

Room tone audio tracks are key for the editor to complete his/her job in editing dialog and generally in making the finished project seamless.

What else can I do?

With the small standalone audio recorder, go and harvest sounds that relate to your film. For instance: if making a film about bee keeping, record many tracks of the bees and their hive.

Set the recorder up next to the hive, turn it on and run. If you are lucky, a bee will land right on one of the mikes, giving you a great up close bee-track an editor would die for. Think outside the interview.

Your editor will love you. Remember, an editor has only what is shot/recorded to work with. Give them options! Record them; have a whole library of sound to use. Think like a filmmaker.

You will not be forgotten.

This is what careers are built on.

POINTS TO REMEMBER

- The DSLR is a powerful tool for making a film, but it has some shortcomings we need to address.
  - Audio input is a 1/8″ unbalanced mike level input.
  - The camera will not power your professional condenser microphones.
  - The most useful professional microphones are universally XLR/balanced cables.
  - Most DSLRs do not allow you to monitor the audio you record.
  - The Automatic Level Control function of your DSLR is problematic.
- Double system can be extremely useful; however, you need to sync sound with picture in post.
- Always monitor your recordings.
- The recording of room tone is critical for post-production.

EXERCISE

Set up your DSLR for a shoot. Leave the room. Have a colleague remove one key element or disconnect a key connection.

Figure it out. Make it work.

Learn how to troubleshoot.

4 THE AUDIO MIXER

It's all about control: ships have the bridge, aircraft have the cockpit, and the sound person has the mixer. With one brilliant device you can control and craft the waves that your microphones gather.

Each mixer is as different as the many dashboards in an automobile, but all allow you to control the same few fundamental functions to produce great audio.

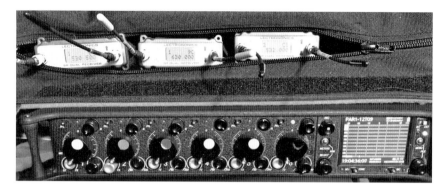

The machine, not the man. A Sound Devices 664 mixer/recorder with three Lectrosonics SR dual channel receivers.

The audio mixer we speak of here is a *tool* not a *person*. It is an electronic device that takes audio, from two inputs to, oh, say 72+ inputs, shapes the sound coming in and sends it various places.

There are two basic types of mixers. The portable field production mixer, and the soundboard/mixing desk/audio production console type of mixer.

The field mixer is small, portable, and pretty explicit in what it does and is expected to do.

For the purposes of this guide, we shall limit our scope to field production mixers, those small, versatile powerhouses that shape and distribute the audio we gather out in the wild world.

SOUND DEVICES 552

Below is a Sound Devices 552 mixer with four Sennheiser wireless microphone receivers.

It also has a nine-person face card to help the mixer identify the cast members wearing wireless transmitters.

The number next to the cast member's name is the frequency assignment the mixer must tune the receiver in order to hear and record the cast member. More on wireless in Chapter 7.

Why even use a mixer? Why can't I just plug a mic into my camera and go shoot?

Well, you can, if it doesn't matter to you what it sounds like.

Of course it matters. That's why we are here.

I have shot with my DSLR and a Sennheiser 416 Short Shotgun Microphone (and a power supply) plugged right into the 1/8″ mini plug input.

Why did this make me nervous?

You can't monitor
Right, **I couldn't monitor** or hear the sound as I was shooting, or watch visual aids (meters), of my recording.

Oh, you mean you want to hear?

You can't control levels as you shoot
If an interview subject suddenly spoke loudly, I could not turn the levels up or down unless I went into the menu, a two-step process that required me to take my attention away from the subject, my frame, and, if I was shooting handheld, would affect the stability of my shot.

You can't alter the audio to make it sound better
My interview, by necessity, was near air conditioners that could not be turned off. My camera lacked the ability to roll off the bass at troublesome lower frequencies.

You are limited to just one microphone
I was limited to just the one shotgun. Awesome mic, but no matter how good your mic is . . . what if two or three people are speaking?

Oh, you want to record more than one person.

See what I mean?

So let's list what a mixer can do:

It provides guidance, both visually with meters and aurally through headphones.

It can shape the nature of the signals you choose to input by altering its sensitivity to frequencies.

It allows you to input mic level inputs using balanced XLR connectors.

It can power your condenser microphones.

It can distribute the signals you mix to various sources at line, mike, and headphone levels.

Some mixers even record audio, with or without timecode.

THE SOUND DEVICES MixPre
The Sound Devices MixPre is a very simple mixer. It has two inputs, LED metering, a headphone output, and two XLR line level outputs. You can switch the inputs (hard pan) left, right, or center, generate tone at

1000 Hz, set the brightness of the LED display, and adjust your headphone level. Headphones allow you to *monitor* with earphones.

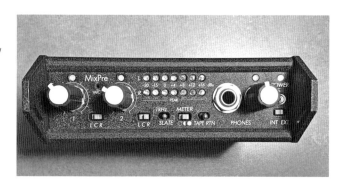

Of all the things this little powerhouse can do, the most important, are two primary functions: *monitoring*. *Monitoring* is achieved either visually by LEDs and aurally through headphones.

MONITORING

When I was first starting out as a sound recordist, I was called to do sound for a film interview on a bluff above the Palisades miles outside New York City. I had just picked up my Nagra IS reel-to-reel tape recorder the day before from routine servicing.

The shot was set, reflectors placed, I powered up the Nagra and everything sounded fine; however, the *modulometer*, or meter used to indicate sound levels, was dead. No movement at all. The little needle just sat there. Naturally, I freaked out a bit. I had been trained to stare at the bouncing needle that told me how

The Nagra IS Modulometer

loud things were, and how far I could go in recording my perfect audio.

What did I do?

I used my ears.

This is the most important resource a sound recordist has. Your ears. I found that just by listening, I could clearly make informed decisions about audio levels to make clearly perfect tracks.

It is important to listen.

Of course, yes, but listening with consistency is super important. What do I mean by that?

First, find a really great pair of headphones, and buy a couple of pairs of them.

They must fit over the ear to isolate sounds other than what you are monitoring from your recorder.

They must have an unbiased frequency response with full representation of the sound spectrum from the lowest bass to the highest highs. Many brands of headphones have a low end boost to enhance styles of music that are popular today.

Not good.

Choose a good, neutral pair of headphones, that are comfortable, rugged, and easy to repair. A system of modular parts, cables, and ear pads that can be easily swapped out if worn out or defective, is a real plus.

Once you have chosen your headphones, stay with them. Do not swap out headphones like pairs of shoes.

Get to know your monitoring source and stay with it.

Audio mixers in recording studios always use a specific pair of monitor speakers to listen to the mix they are building. They can count on these reference speakers because they have spent years listening to their work and the speakers have become a constant reference for them.

Find a pair of good headphones with clean unbiased sound and stick with them.

Monitor levels on your mixer or recorder (the headphone level) must be loud but not to the point of ear damage. Keep this level constant.

Your ears are your most important tool!

Watching the needles or the flashing lights

This deck has these flickering LEDs. Or these bouncing LCDs, or these little gauges . . . what do they mean?

That's your incoming signal.

They indicate the levels of energy that are being fed into the inputs of your mixer/recorder, or rather the differences between the highest and lowest acceptable levels of those input signals.

You typically set the highest levels below the maximum point of what your equipment can handle, and keep the quietest signals well above the noise floor.

Noisy floor?

Noise floor. All electrical recording devices can be heard working if you turn up the volume during playback.

This is known as **signal-to-noise ratio or S/N.**

A **high signal-to-noise ratio** means the sound you record is much louder than the noise floor of the electronics at work to produce the track.

A **low signal-to-noise ratio** means that the cheap/defective components in the record chain are almost as loud as the incoming audio signal making your tracks full of hiss and noise and sound, basically, like crap.

What happens if I mix a sound too loud, and the lights/needles go into the red?

Your ears will tell you if you go too far, or not far enough, but meters can help you set levels much quicker and easier, warning you of potential problems before they occur.

All metering systems have a 0 dB point clearly marked on them.

That is the border between safety and toward distortion/dropout.

If you play it safe and record well below 0 dB your recording, when monitored at normal volume, may have a noise floor (hiss) that results in noisy tracks.

That's bad.

There are reasons why one microphone or mixer is much more expensive than another. Just Listen. The pricey gear is silent. The budget gear has that hiss. **Low signal-to-noise ratio!**

What am I to do?

If you record audio too low, you risk an unfavorably low S/N (**signal-to-noise**) ratio.

Record your levels too hot, at and beyond 0 dB, and you risk distortion or dropout.

You have to walk a fine line.

That's why you are there, to ride levels and make a great recording.

Recording an analog signal, on tape, say (yes, some people still record on tape), that is too hot, well above 0 dB, say +12 dB, you can hear the signal begin to gradually distort, allowing a quick reduction in levels, backing away from the cliff, shall we say.

Digital systems, however, have no headroom above the maximum audio level. Exceeding the maximum audio level will cause dropout: a total loss of signal. You just fell off the cliff.

Because of this, headroom is created by choosing a "zero" point well back from this "sonic cliff." Digital meters are set depending on the standard. Across the globe standards have set 0 dB at −18 dB to −20 dB. Can you record above −18 dB?

Of course, just beware of the risks.

Watching the meters
Let's look at what and how they indicate.

VU meters
This is the granddad of all level indicators. VU stands for Volume Unit and has been used since the beginning of radio, broadcasting, and recording. It was invented by Bell Labs for NBC and CBS for measuring and standardizing the levels of telephone lines. It is designed to show an average of the voltage

VU meter

level of incoming electrical signals, something referred to as RMS. RMS (Root Means Square) is actually an engineering term meaning a kind of "fudge" or "fuzzy" standard of measurement. A constant tone will be accurately measured, but anything more complex will be a lot less accurate. VU meters intentionally slow response times to average out the incoming signals. Often devices using this standard have analog needles which are moved mechanically, adding a physical delay to the average reading of your audio.

Peak Programme Meters (PPMs)

The Peak Programme Meter (PPM) is a more advanced and elaborate system with a much faster speed of response, allowing you to drive audio louder, giving better signal-to-noise performance. PPMs are usually LED (light emitting diodes); however, some devices have the old-school swinging needles.

Some devices have combined the two with VU response shown as a solid line and PP indicated by a floating dot just ahead of the line. The classic Sound Devices 442 and 552 Mixer as well as the 633 and 664 mixer/recorders have this type of combination as a menu option.

Insurance: the limiter

The basic aim of any limiter is to automatically suppress the input level when the audio signal exceeds a certain threshold. Most field production mixers have a limiter circuit due to the unpredictability of the type of sounds we record. In the unpredictable real world, ambient sounds, the voice of a non-professional interview subject, or unexpected events, can all threaten your tracks with loud sounds which cause overload and drop out.

All limiters are not created equal. You can barely hear really good limiters working. Some not-so-good limiters you can hear "pumping" or "punching holes" in the sound as the circuit attacks the loud signal and releases the attenuation back to 0 dB.

Get to know your limiter and its, uh, limitations.

Bass attenuation

What does attenuation mean?

Attenuation is the gradual reduction of intensity of waves—in our case, sound waves.

Most mixers have a control that attenuates bass.

What does that say next to the switch?

The photo above indicates: (−) minus bass dB at 100 Hz:

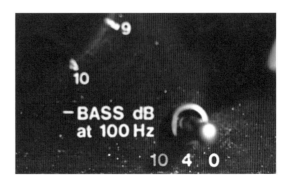

- The switch at zero is inactive.

- The switch at 4 means it eliminates 4 dB at 100 Hz.

- The switch at 10 eliminates 10 dB at 100 Hz.

The measure 100 Hz is the heart of low-frequency mass in the audible frequency range. Remember, we are able to hear from 30/50 Hz on the low end through 15 kHz/20 kHz on the high end.

Why are we concerned with just bass?

If you think about the forces that can alter our tracks and make them less awesome, there is no more powerful force than the wave energy in the lower and longer frequency ranges. They tend to muddy up our voice recording, and reduce clarity.

What are the situations where bass energy can affect our recordings?

- If you have a lavaliere mounted to a person's chest, or under clothing, that cavity can be full of lower frequency energy that could make the recording sound pretty unnatural. To make it sound a little better, *attenuate* the bass.

- A particular room may be excessively reverberant or "boomy." This reverberation could be caused by sound waves bouncing off various surfaces more than a few times and returning to the microphone, off axis, in the lower frequencies. Roll off some bass to make it sound better.

- A recording done in an urban setting with the everyday rumble of traffic or trains nearby could make for a distracting recording. Roll off the bottom end.

- The wind itself is nothing but giant bass waves. If you are in a situation where wind is audible on the mike, try rolling off some bass.

Caution: Once you have made a decision in the field to alter the nature of your recorded tracks, it may be impossible or too expensive to restore that which your decision has affected.

The overuse of bass attenuation can make the human voice sound thin and "tinny." Be sure before you roll off the bottom end that you yourself are not being affected physically by the same bass that you are monitoring. In other words, because you are feeling the same bass you are hearing, you may misjudge its intensity.

Never make decisions in the field that cannot be undone in post.

Mic level and Line level

The MicPre two main XLR outputs audio at only <u>line</u> level.

As we learned in Chapter 2, sound waves that are recorded by a dynamic microphone produce a weak electrical signal. This signal must be amplified to a level which can be more easily manipulated by other devices such as mixers or recorders. This amplification is done with a preamplifier (AKA preamp) which bumps up the signal to a useful level.

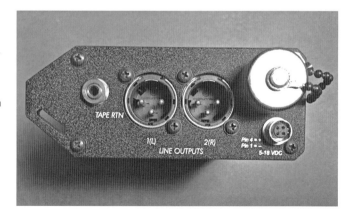

If your correspondent is speaking into a handheld dynamic mike, and this in turn is running into say, Input One of your mixer, the XLR(F) input will usually have a switch nearby allowing you to set **line** or **mic**. Clearly, this should be set at **mic** level.

You are using a **microphone**, no?

But there's a line attached to the microphone.

True, but line exists to deal an input that is more powerful than the faint hand full of millivolts output coming in from your dynamic microphone.

Coming from where?

Let's say you are recording a friend's band as they play at a club. You set up your recorder and get a feed from the mixing board which controls the house speakers and the monitors for the band. The feed is an XLR-XLR cable you can plug right into your recorder. Great!

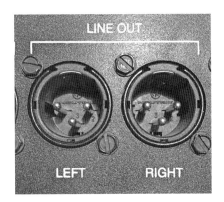

If your recorder is set for your handheld dynamic mike (mic level) that you have been using for your podcast, *you have a problem*. The incoming signal from the board is too powerful. It will overload your device. It won't be happy, you won't be happy, your tracks will be all distorted. Too much level!

You must set your recorder to **line level**. After all, you are receiving a signal not from a **microphone** but from a **line-level output**.

By switching the input to **line**, you are *resisting* the incoming signal.

Let's look at it this way. *Really* look at it, by using the *eye* as an example.

The eye?

Yes, the eye. OK, you know how when you are in a dark, or dimly lit room, the iris of your eye opens as wide as possible to let in any available light so you can see? It's wide open.

OK, got that . . .

Then you step outside into the sunlight. Wow. Your eye's iris closes down to keep from overloading the photoreceptors in the back of your eye.

The iris *attenuates* the incoming light. It *resists* the amount of light coming in.

With your recorder it is very similar.

At **mic level** your recorder is **open wide** for all incoming information.

At **line level** it is **closed down**, very much like your eye in bright sun, to *resist* the amount of light coming in to photoreceptors in the back of the eye.

We are able to measure this *resistance*. We also have a term for this *resistance*.

This resistance we measure in **Ohms.**

Ohms are indicated by the symbol for Omega.

Georg Ohm found that there is a direct proportionality between the potential difference (voltage) applied across a conductor and the resultant electric current. His writings and research from the middle of the nineteenth century established the standards for electrical resistance we use today.

Mic level is represented in units of resistance, or **Ohms**. **Mic** level is commonly **150 Ohms**.

Line level is commonly **600 Ohms** of resistance.

*What happens if you send **line out** to the camera which has its inputs set to **mic in**?*

Georg Simon Ohm

Terrible distortion will occur creating unusable tracks. Unusable tracks mean never getting hired again, or a ruined project.

This is a really confusing situation for the fledgling recordist, for on occasion, seemingly you have done everything correctly, but you hear *nothing*.

The **mic/line** switch may be set in the wrong position. You have your microphone plugged into an input set for **line level**. It resists the incoming signal so much you hear nothing.

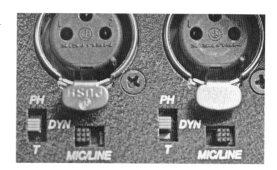

The **mic/line** switch is often positioned on the side of the portable mixer next to the individual XLR(F) input, buried in the nylon flaps of the case, sometimes even beneath the XLR connector, and if that weren't enough, the current miniature designs of portable mixers can make this switch physically quite small.

Next to the XLR(F) inputs are switches for mic powering or not, and mic/line level attenuation. Tiny but important switches.

The switches for microphone powering are to the right of the mic/line switch.

Direct/Return

Most mixers have a switch which allows you to listen to the audio you are processing in the mixer itself, or the *return* audio from the camera or record device. If you are using a multi-pin snake input cable which includes a monitor line, you can plug into the headphone output of the camera or deck and hear what it is receiving.

Direct/return switch on a Sound Devices 442. Midway between A and B is direct.

This is important! *Always monitor return!* This will confirm that you are sending excellent tracks, and that the cable you are using to send these tracks is functioning properly, and the recording device (camera/deck) is set properly to receive them with the correct levels and impedance.

Here's a story: a production company travels to an exotic island. To save money, they hire a local sound man. They shoot for a week, with talent doing pieces to camera about how amazing this island life can be. On returning to New York, it is discovered that the audio is distorted and unusable.

Disaster. How did this happen?

The local sound man, lacking experience, sent a **line level** signal to the camera which was set to **mic level**.

When he sent tone to camera, the levels were both looking good at −20 dB. The tone through his headphones *from the mixer* sounded normal.

He was listening to **direct**. He never listened to, or monitored what the camera was receiving from his mixer. Had he switched to monitor the camera audio output, he would have heard a distorted blast of something like tone. He never went there. The cameraman *never monitored audio* from the camera. You are a team. Your cameraman and yourself, the sound mixer. Since most video cameras are the audio recording device as well as the image recorder it is a really good idea that the cameraman listen to the audio as well. He listens for quality control, for dropouts if a wireless hops system is in use (more on this in Chapter 7), and generally takes mutual responsibility for the work you are both doing. In this case the cameraman never monitored takes, or checked playback for audio quality. Perhaps working in paradise might have had something to do with the attitude of the crew. Being too ready to enjoy the tropics perhaps made him neglect the usual checks and balances.

Always check playback, no matter what might be going on to distract you. Distractions: bad weather, cold, rain, heat, parties, nightlife, drink, sex . . . all tempt you to be preoccupied with personal comfort and let things slide. Stick to your procedures and make sure you have the shot. This mixer was listening to **direct** for a week and everything sounded great, and no one caught the distorted mistake.

All good. **No good.**

Powering microphones

As we learned in Chapter 2, the tribe of microphones has two groups: those that need power to function and those that work just fine without it.

Your mixer will allow you to use both kinds to the best of their ability.

Dynamic microphone

Again, as we learned in Chapter 2, a dynamic mike is simple and explicit, plug it in and with the powering switch set on DYN (for dynamic), go forth and record sound.

The DYN switch setting means that no power is being sent to the mike, for it generates its own signal.

Caution: Some lesser mixers add noise to your fine recorded tracks if phantom powering is switched on when using dynamic mikes.

Condenser microphone

If you have selected a *condenser microphone* such as a Sennheiser 416 Shotgun, you must find the mic powering switch on the mixer and select *48v Phantom*.

Phantom?

It sounds mysterious, but actually is a term indicating that the device your mike is plugged into is sending DC current through the cable, to silently power that condenser mike.

There are two systems for conveying power to microphones via the mike cable.

One is phantom power (12–48v). This is the standard today for powering condenser microphones.

The other system is T Power (12v).

T Power is a legacy technology from the days when the primary image recording device was a 35/16mm camera and the audio recorder was a reel-to-reel analog tape recorder made by Nagra.

These machines and their inventor, Stefan Kudelski, are legendary.

Our simple MixPre allows us three options when it comes to mike powering.

- top position: 48v Phantom

- mid position: Dynamic

- bottom position: 15v Phantom.

The switch to the right of the powering selector allows you to roll off bass—or not, if you select the middle switch position.

The Nagra III

The top rolls off bass to 80 Hz.

The lower position rolls off bass to 160 Hz.

The final switch is a limiter.

The small recessed pots allow you to set the limiter attack threshold, or how aggressively the limiter will do its job.

The limiter switch is off in the middle position,

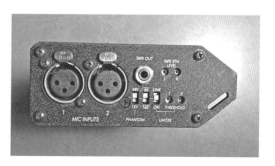

On in the bottom setting or links both input 1 and 2 in a stereo pair.

The recessed controls to the right of the headphone jack allow you to set levels of tape return volume.

Turn on your mixer, and plug everything in.

There's no sound. I hear nothing.

Troubleshoot your problem.

Systematically trace your signal flow.

I find it useful to trace from *out to in.*

By this I mean: No audio? Trace the signal flow from the microphone, to the boom cable to the input select switch, to the pan pot to the . . . and so on until you have found the answer.

Do you see levels on your meters? No?

Mike connected to cable.

Mic cable plugged into correct input.

Correct input level turned up.

Headphones: Connected and volume up.

Input set to the correct impedence:

Ah, it was on LINE, not MIKE.

It is usually something minor—a switch in the wrong position, a cable in the wrong input, be methodical and trace the fault.

OK, everything works, hook up to the camera and send tone.

What is tone?

The tone I'm talking about is a signal, typically 1000 Hz, generated internally from your mixer.

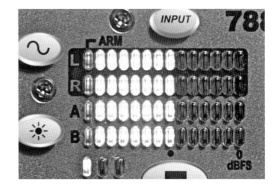

Meters indicating tone on a Sound Devices 788

The other type of tone, *room tone*, we discussed in Chapter 3, and will further discuss in Chapter 12.

Why tone? This tone creates a reference signal which when recorded onto media allows the technicians in post-production to set levels on their editing equipment.

It is also a very useful tool to set levels in the field on the camera/deck you are sending audio to. Send tone, set your levels, and record up to those levels, and everything will be just fine.

Pads/attenuation

A really useful function you find on most mixers is the ability to *pad or attenuate* the incoming signal.

What does this mean?

Again it is our friends the Ohm brothers . . . electrical resistance.

Broadly, in electrical engineering and telecommunications, attenuation *affects* the propagation of waves and signals in electrical circuits, optical fibers, and in the atmosphere (radio waves).

Your mixer uses electrical resistance to *reduce* the amplitude of the incoming signal.

How do I use this? When do I use this?

OK, you have to take a feed at the Glass House in New Orleans where the Dirty Dozen Brass Band is playing.

You run an XLR cable from your mixer to the output of the house mixing board. With headphones on, you listen and the music is loud, bordering on distortion. You just touch the knob on your mixer, and it pegs the meters right away.

Way too much signal.

What do I do?

You attenuate or "pad" down the incoming signal.

You can do this two ways:

Use a Line Input Adaptor which is a barrel that has an XLR(F) at one end and an XLR(M) at the other. In the barrel are resistors that pad the passing signal down 50 dB.

Or if your mixer is so equipped, Switch in −10 dB of attenuation. Better? No. Still too hot? Try −20 dB of attenuation. That's even better. Turn the flow down. Attenuate it. Generally speaking, the phrase "padding it down" is also applicable for what you just did.

Multi-track and stereo recording

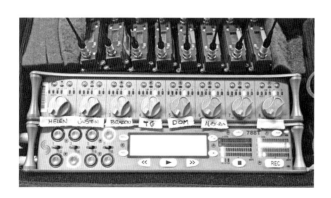

True stereo recording, an art and science unto itself, will be dealt with in Chapter 13.

First, let's deal with the ability of your mixer to assign tracks.

Production Sound Mixing

Most mixers allow you to hard pan or switch a track to be sent to either the right or left output or the mixer.

It is often important to isolate subjects in an interview. The mixer below has its hard pan switches in a row below the pan pots.

Pan pots allow you to assign by degrees which channel is designated to right or left.

Other mixers combine these functions in one switch, illustrated by the row of switches below the soft pan knobs.

The example above has the pan pot switched to full right channel.

What is a pan pot?

Pan pot is a term used by recording engineers to refer to a *pan control.* A pot is a slang term for *potentiometer*, a simple device that provides variable resistance. The most common potentiometer is

an analog volume control, like the volume knob on a radio. Unlike the usual usage of a potentiometer, the *pan pot* or *pan control* distributes its source sounds with constant power so that the overall signal power level is constant regardless of the knob position.

Why assign either right or left to a track?

We assign tracks to separate them from each other.

Sound recording for decades combined everything into one Monoral or mono track.

In the 1960s, film production began to employ two track, then multi-track recorders. Combining everything into one track limits flexibility, and exposes your recording to great risk.

Say you have two subjects on wireless mikes. A burst of static, a hit, happens on one transmitter. If it's a mono recording, where both tracks are centered, it will be heard on both. If your tracks are separated, one right and one left, you have at least one clean track.

You limit your exposure to risk by isolating your tracks.

In post-production, the editor can change levels, insert or subtract, and change the timing of the spoken word. In other words, the editor has a much greater flexibility with separate tracks.

In a multi-camera interview, all cameras can be fed exactly the same audio to specific channels. Say, the interviewer on channel one, and subject on channel two.

Mixer custom features

Pre-Fade (PFL) ability: This switch allows you to monitor the *input signal* before you actually activate the channel to be recorded. (By turning up the knob/fader into your mix.) The ability for a mixer to monitor what is being said before fading in a track is a great tool. The audio you hear is only present in the monitoring circuit and is not sent into the mix or recorded. Typically this switch resides above each channel fader.

Why is this useful?

Let's say you are mixing five wireless mikes during the shooting of an art gallery opening. There is one cameraman you are following and getting great tracks for.

You notice one of the subjects you have wired in a corner, whispering to an art dealer. Although you are recording a conversation right in front of you, you want to hear this other conversation to determine if it is useful for the film.

Hit the PFL switch on the corner whisperer's particular wireless receiver's input channel. It overrides all other signals, in the monitoring circuit only, and allows you to isolate and listen to that individual conversation. You hear an important conversation going on. Tap the cameraman on the shoulder and direct him to the corner.

You just scored some content.

Slate mike

This is a small microphone on the front surface of the mixer which allows the recordist to verbally slate tracks or communicate with a boom operator, utility, or other member of the sound team. *Do not hit this switch during recording.* It will be heard/recorded on all tracks.

On this mixer, pre-fade is indicated by PFL

M/S stereo ganging

This switch allows two inputs to be controlled by one fader, something that makes controlling M/S stereo recording much easier.

M/S stereo will be discussed in Chapter 14.

Recording ability

Many mixers have the added ability to record audio as well as functioning as a mixer. On some, this is a non-timecode recording on an SD card for transcription purposes. On others, it is a full-blown recording capability on a CF card or hard drive with timecode.

More about timecode in Chapter 11.

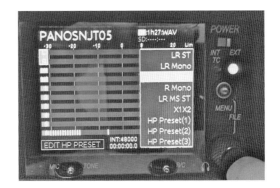

Sound Devices 664 monitor menu

Headphone monitoring options

Some advanced mixers allow numerous monitoring options (what you hear through your headphones). This is huge in keeping track of your signal flow, and really just knowing what's going on in your mix.

- A mixer accepts various inputs, either microphones or a line feed from a separate source. With a mixer we can control the levels and shape the sound we accept.

- With a mixer we can monitor not only the signal coming in from our microphones and sources, but we can also listen to the return signal from out record device.

- The mixer can determine which way we send our signals and to which track or device.

Connecting to camera or deck checklist

The connections to the camera/deck are the same as your output.

- Your output impedance values are the same for each channel: right/left is either line or mic. (Both must be the same.)

- Monitor line is connected to the headphone output.

- Monitor volume is increased to a usable level.

- Monitor select switch is set to your preference (if applicable).

- Send tone.

- Monitored return is **loud and clear**.

- After setting levels with tone **listen to a non-tone sound input like a microphone**.

- If feeding multiple cameras, check for **channel conformity**. Right is right on all cameras, left is left, etc . . .

- When shooting begins: **monitor return, but check all your individual feeds for clothing rustle, battery level, bass incursion . . .**

And remember, the camera is your record deck; you must monitor all the time . . . .

Semper Vidant: always be vigilant!

EXERCISE

Set up a mixer in a typical one-camera shoot situation. Input a boom and a lavaliere into channels one and two.

Run tone, set levels, make sure everything sounds great.

Leave the room.

Have a colleague change three settings from the following list:

- Set mike powering: dynamic, phantom.

- Set input level: line or mike.

- Turn down headphone level to zero.

- Remove the power from the mixer (take out the batteries or just turn it off completely).

- Disconnect the return from the camera.

- Disconnect the XLR cable from the mike.

- Disconnect the headphone jack and tuck it under the mixer.

Come back in and troubleshoot.

Repeat the exercise three times.

To add extra anxiety time, limit each troubleshooting session to five minutes.

5 CHOOSING A LOCATION

Locations: involving audio perception in the choice of where to shoot.

"It's harder to make real audio than special effects audio."

Alejandro Gonzalez Inarritu

One of the most important things to consider on the road to good audio does not involve microphones, or digital recorders, or any batteries. You just need the tools you were born with: your ears and your voice. Your ears listen, and your voice critiques the choices of where the shoot is to take place.

HOW TO FIND THE BEST SPACE TO RECORD SOUND

More often than not, there is no choice; you just have to make it work . . .

Once you become a sound mixer you will never be the same again.

Why?

You become *aware of sound* . . . everywhere you go.

Do you ever walk into a room and feel mildly annoyed?

It may be the sound . . .

Usually offices, often indifferently maintained civic spaces, are lit by rows of fluorescent tubes.

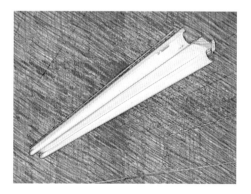

A fluorescent light tube has a lot going on inside. Electrons are striking atoms of mercury vapor, which emit rays of ultraviolet light, and those rays strike a phosphor coating which are energized and emit white light. All this action causes the tube to hum, especially when it has been in service for some time.

If a new tube is installed next to an older one, the high-frequency sound waves emitted not only change constantly but clash sonically. Hanging around in this sound field for any length of time can really be annoying. Couple this with hard surfaces, glass, or tile, which reflect this sound and further change its tonal structure . . . and it can make you just want to run screaming down the hall.

Very few people understand this annoyance. Even fewer people are aware of the sound of air-conditioners, water coolers, refrigerators, clocks, and other appliances that intrude on our everyday life experience and prevent us from recording perfect tracks.

In mainstream TV and feature film production there are a group of experts called *location scouts*, who know a particular area/city so intimately that if a director asks for a certain look—be it a house, garage, park, street, or storefront—they know exactly which places in the general area to present as possible shooting locations. If they don't, they will scout for them. Unfortunately, rarely does their expertise include the sonic world.

This can be a challenge to the sound recordist.

A location may be chosen, usually by the director, with the aid of information of the scout, without enough concern being given to the surrounding sound environment. The level of external sound, just like the moving shadows of the sun, will change throughout the day. Before 7 a.m. it is quiet, then the deliveries begin, people wake and open shops, transportation picks up in activity. There may be a lull between 9:30 a.m. and 11:30 a.m. and then the lunch rush. Noon often brings church bells or a blast on a siren.

School's out at 3 p.m. with all the attending activity. You have an hour or so and then the evening rush begins to ramp up.

And so on. You can see that noise levels rise as activity increases.

Production people may scout during times that little ambient noise is occurring, say on a weekend when normal work surrounding the location is not taking place, or those scouting are just oblivious to sounds that can make life difficult for the recordist. The producer should take an active interest in locations that are suitable for all departments, for dialog that is compromised by impossible ambient noise will have to be re-recorded (*ADR: Automated Dialogue Replacement or Additional Dialogue Recording*) in a studio at great cost.

ADR and the man that made it feasible, Otto Popelka won an Oscar in 1967 with an electronically controlled looping system he developed at Magna-Tech.

In documentary production, you must shoot in the place where an event is happening, or has happened, usually meaning you have little or no choice.

Years ago, while working on the series *Dancing In The Streets* for the BBC, I was hired to record an interview with Lou Reed of the Velvet Underground. The location chosen was Gleason's Gym in Brooklyn. It is a boxing gym. A big open space. Somehow, during the scout (or *recce*, as the English say), which I was not present for, no one noticed, or was informed, that a printing plant was in residence just upstairs. The plant's offset web presses, each the size of an airport shuttle bus, were running at full speed, so present, so loud, that the hanging lighting fixtures in the gym were shaking back and forth.

Needless to say, I was not happy. Especially when the producer requested that should Mr. Reed ask about the loud ambient sound, I should tell him it was fine.

I did not respond favorably to this request. I thought, but did not say (key to being rehired), that I would not lie to someone who has provided so much of the soundtrack to my life.

Mr. Reed did ask. I told him it was pretty terrible. He sighed and said something to the effect of: ". . . in a perfect world . . ."

Years later, I watched the interview. The throb of the printing presses was clearly heard even above the bed of music mixed below the interview.

QUESTIONS FOR THE PRODUCER OR PRODUCTION MANAGER ABOUT THE LOCATION

Exterior

- What time is the shoot? This is most important. If an interview starts at 4 p.m. and runs for an hour and a half, you will find in urban environments that rush-hour will start and things will become problematic (noisy).

- Is a highway/street/airport/firehouse/hospital nearby? (traffic noise, aircraft overhead, sirens)

- Is there any work in the general vicinity? Like construction or street repair? (jackhammers)

- If a rooftop or beach, how are the winds? If a beach, is the surf going to be in shot? (it's loud)

- If it rains, what is the composition of the roof directly above the location? (drum, drum of the raindrops)

Interior

- What is the physical composition of the space? Floor and wall surfaces: tile or carpet?

- Are there windows that look out onto a possibly noisy street or similar situation?

- Can you control or turn off the HVAC (air conditioning system)?

- Is there work in the surrounding offices/rooms adjacent to the location?

- Is the ceiling soundproof? Can you hear those high heels cross and cross again as the workers do their business?

- If it is a bar or restaurant, can the ice maker/machine and refrigerator be turned off?

- Are there pets/children around?

Refrigerators. Documentaries often shoot where people are most comfortable. That means a lot of kitchen shooting. What is usually in a kitchen? Besides a lot of reflective hard surfaces? The fridge with its 50/60 Hz hum. As a sound recordist it is your job not only to turn it off, but also to *remember to turn it back on.*

I stick a piece of tape with REFR written on it to my mixer or sound bag.

It's a bad thing to ruin some family's entire load of food just because you are absent-minded.

Unplug/but always replug

It's so easy to roam a location and unplug ice machines, freezers, and vending machines when the lighting crew is taking for ever to get that look just right.

However, at the end of shooting, things tend to move faster, tasks pile up, and fatigue can take its toll.

Always remember to leave the room as you left it.

I find that sound blankets are useful. You either need one or ten. Usually nothing in between.

A sound blanket is the same thing as a mover's blanket or furniture blanket. It's a big padded blanket usually 6 by 8 feet used to wrap pianos and such to protect them in shipment.

A lot of getting the best deal for your craft involves diplomacy. If you are there right at the beginning of the decision-making process, as the cameraperson and producer choose a spot inside a room for a certain

look, be there reminding them of what is best for sound. Not demanding, but reminding them that we are a team and what is best for the production. It is another thing if you speak up too late and then everything that is set up has to be moved, re-lit, etc.

How to survey a location

If entering a location for the first time, just stop for a second and listen. Get everyone to be quiet. What do you hear?

One thing that I do is to walk into the middle of the interview location and clap my hands together.

Then I listen for echo reverberation or "liveness" of the room.

What is the room constructed of?

Walls: plaster, sheetrock, wood paneling, brick are the usual, not so bad for reverberance. But tile, or concrete, steel, or glass, can be a nightmare.

Walls are important, for they are in the shot.

Floors, unless the shot sees the whole world, can be covered or blanketed. They are often too low to be in shot.

What do the floors consist of?

New floors covered with carpet? Great.

Old bare wood floors that bounce sound and creak under you even if you are stand still?

Disaster.

See everything. Notice everything.

Sometimes problems do not make themselves known until you are shooting.

I was mixing an *American Playhouse* in a restored mansion in New Orleans. We were outside in a narrow courtyard at the center of the structure. It had been drizzling, and plastic sheeting was hung over the courtyard so we could work and stay dry underneath. The shot was a long dolly move in front of the two main actors who walked the length of the courtyard. We were set, and did a rehearsal. The movement of the dolly, its crew, my boom operator, and the actors displaced so much air under the plastic that it actually rolled like a wave breaking over a beach, making a rumbling noise like distant thunder. Surprise.

After cutting some slits in the plastic, it settled down enough when we moved for us to do a proper sound take.

What surrounds the building that might affect the interior sound field?

- Trains. If they're around, how often do they run?

- Aircraft. Where is the location in relationship to the airport? Check the map. Check the runway heading. This is important. Airplanes land into the wind. On the scout day the nearby runway may not be in use. On a shoot day, they may be rolling in every three minutes.

- Yard work, leaf blowers, lawn mowers. If they're around, when do they start and/or finish?

- Proximity of the location to a fire station/hospital (sirens).

- Street traffic. The sound will change dramatically if you start work in the relative quiet of mid-afternoon and shoot through rush hour.

- Garbage collection, backup alarms, other truck generated noise. If you are shooting near a warehouse, you will be hearing backup alarms all day.

- Church bells or noon/hourly chimes may haunt your shoot.

- Kids coming home from school.

- Animal life: barking dogs, birds, crickets, cicadids.

What surrounds the room that will affect the interior sound field?

- Exterior hallways. Can you lock it up? (Have production make signs.)

- Elevators chiming in the hallway. Can they be disabled?

- Nearby office workers. Are any of them wearing loud shoes/heels?

What can you do to make the room itself sound better?

- All phones off: personal/cell and office/desk and intercoms.

- The most important is the HVAC system. Can you turn it off? No? You can control the HVAC even without turning it off. To reduce the sound of air being transported about the space, you can place cardboard, blankets, books, even equipment cases on top of offending vents. It may not be pretty, but it matters not if it's out of the shot.

- What about other surfaces not being seen in the shot? The floor can be blanketed, or covered with available material to knock down its reverberance.

- Windows can be closed or covered.

- If visually acceptable, the interview subject can be angled within the room so as not to face flat surfaces squarely, which would cause bouncy acoustic reverberations.

When recording tracks in a room, particularly a small one, standing waves can occur (Chapter 1), especially at low frequencies. The room will sound oddly compressed, like your head is in a cardboard box.

How do I get rid of this?

You can move the interview subject within the space, or you can place absorbent materials in the corners of the room where walls meet, or in the place where the floor meets the wall itself. Some well-placed sound absorbing material, such as blankets, cushions, or coats, will break the chain of standing waves.

In a perfect world, with no limit on time, resources or cooperation, you can employ all these strategies. In the real world, you have to live with many irritable forces. Often a fan sound or hum from a computer is present. If a sound is constant, you can usually tolerate its presence on the soundtrack. If it's intermittent, it is much more difficult to overcome in editing.

In documentaries, a recordist just has to make do with the context of the film. I did two films about Air Force One, one for the History Channel and one for Discovery. There is a constant roar even in a sophisticated aircraft like the modified 747 the president rides in.

In active sound fields like aboard an aircraft, or a factory floor, nothing beats a lavaliere/wireless combination. Especially if you can position the mike as close to the speaker's mouth as possible.

On Air Force One, we were roaming all over the plane shooting the crew as they performed their duties. Wiring each person with a transmitter and lavaliere was impractical. Most of my production sound was gathered using a Schoeps CMC 4 and a Mk 41 capsule. The capsule alone was at the end of a lightweight boom connected to the CMC 4 via the Colette Cable. The ceilings in many places in the aircraft are quite low, especially in the cockpit. The low profile of the capsule/Cette combo allowed me to squeeze into really small places.

In the context of the film, a level of background noise is acceptable, for the audience knows we are on an airplane. As long as the spoken words are crisply audible, a given level of background noise is fine. Remember, it must sound *natural*.

POINTS TO REMEMBER

- Be aware of your surroundings and the implications on the quality of your work. You may be the only one listening.

- Just as the sun moves across the sky changing the light that surrounds you, so does the sound scape as the day changes rhythms.

- Always try and be part of a location scout.

- Always be aware of the time of day production plans to shoot.

EXERCISE

Break up individually and in the general area, find the worst place to record sound and then the best place. One each for both interior and exterior.

Write each of the places on an index card.

Group back together, and combine all cards into a deck and shuffle it.

Have each of your colleagues pick two cards.

Go to each card location and explain the strategies for shooting there.

6 BOOMING:
The View From Above

To be a great sound person, one has to understand the art of booming. If you are holding the microphone boom, you must be the most perceptive person on location.

The full understanding of the capabilities of the microphone you hold, the shot as framed, the possible shadows or reflections that may cause problems, and the ability to move in a space are all required to succeed.

An RCA KU2A "Skunk" boom microphone (1930)

HOW TO GET THE MOST NATURAL SOUND IMAGINABLE (USUALLY AGAINST ALL ODDS)

To explain the process in the simplest terms:

A sound person must hold a microphone as close as possible to a person speaking but still out of the camera's frame. That microphone is attached to a collapsible telescoping pole.

The reason for employing such a device is obvious: the closer the microphone to the person speaking, the better and more natural it will sound. However, camera frames can be pretty wide, but are usually wider on the sides than on the top. Held at a right angle, just above the frame the microphone is directed to follow the speaker, on axis, for the duration of the shot.

You can boom an interview.

You can boom a scene of a fictional narrative film.

Another scenario is during the shooting of a documentary to boom during an actual live, happening, life event. This is pretty challenging. You have no control over who speaks when, what the camera frame size is, or if or when it might turn in your direction.

You boom during the production of a documentary in uncontrolled situations because that is the only way to get sound. After all, the population of the whole world does not come pre-wired with lavalieres and transmitters.

To be a good documentary sound person, you must have great *situational awareness*.

Situational awareness means being conscious of two worlds:

1. First there is the scene you are filming. In the technical sense, what is happening, where your mike is in relationship to the frame, where the cameraperson or scene may go next.

    You must know the camera and what its concerns are: What is it seeing? Where is it going? It's a challenge to stay as close to the subject speaking with the mike, but stay out of the frame.

2. It is essential to be aware of what is going on in real life within the room, around you and your crew bubble of film production. Is something going to happen that will suddenly change the scene you are filming? Are the subjects going outside? Is someone about to enter the room?

Who is going to speak next?

This is key. In a group dynamic, it is your job to allow the film to hear everyone.

People don't usually just start speaking. They kind of wind up before they utter a word. They sit up straighter, raise their head higher; really, most people present a whole body language of pre-speech. You have to watch for this. If you have just met them it's tough, but once you have spent an hour with them, you will learn their signs.

Once they begin to speak, you get there with the boom. Usually you will beat the camera to the speaker who will wait for a smooth way to transition. Even if camera zips right to the speaker, they still have to focus, frame get set . . . You have beaten the camera by a mile.

Listening is pretty important. Not only listening to the content of what is being said, but the rhythm of the sentence.

If a speaker hesitates, or slows in their presentation, often a second or third speaker will jump in. Be alert and don't "sit" on one person without being ready to move.

It is often useful when booming a group, or even a pair of people, to hover in the middle space between them until one of the group seizes the initiative, and begins to make a point. Then creep in and get on axis.

If a comment is made by another on the topic your subject is speaking on, you can briefly feather the mike to that speaker, stay with the main speaker until the end of their statement. It cannot be stressed how important it is to listen and anticipate who might speak next.

Always be ready for action, ready to roll, for real life has no call time.

Even when you are not shooting, be ready to roll. Don't take all your gear off and leave it in a pile across the room. Keep it close.

Don't necessarily cut when it seems the moment has ended. Keep rolling a little while. Even if the camera has cut. Some people relax and say useful things when they feel they are no longer "on." Keep your finger on the trigger to record vital information even when camera is down.

Remember in a documentary, sound leads camera in presenting information. Nothing is more powerful than words in the voice of an interview subject.

On the truth scale, it totally trumps voice-over narration.

It may seem obvious, but *you must always have your boom handy*. On wireless-heavy productions many times a mixer will clip it on to their mixer bag, or even leave it somewhere else during an interview.

This is really lazy. What if the wireless transmitter dies? What about sudden clothing noise? Boom and Lav: one on each channel is the way to go.

A good documentary sound person must have great instincts, peripheral vision, and the undefinable ability to become invisible. You have to blend into the woodwork so as not to influence the prevailing mood of the subjects with your presence.

Oftentimes the presence of the camera and its all-seeing lens is distraction enough for the sound person to fade into the scenery. It also helps to be a great pool player, and be a good dancer.

Billiards or pool helps with figuring out the geometry equation of room plus length of boom divided by the movement of talent plus focal length of the camera.

A good dancer?

You are the cameraman's dance partner. You must shadow his every move. If the cameraperson backs up suddenly, you have to bounce backward to avoid a collision. If they walk backward in front of a walking, talking subject, you must boom and stay on axis, walk sideways/backward keep a step ahead, look out for obstacles the cameraperson may encounter possibly by guiding them with one hand on their belt or cue them with gentle prods left and right.

Sound like fun? It is hilarious, and exhilarating when it works out and you get an amazing scene.

I was recordist on an *American Masters* profiling the singer Paul Simon. The cameraman, Christian Blackwood, and I were in front of Mr. Simon and Neguinho do Samba, the director of the street-percussion tribe called Olodum. We were shooting these two musicians walking down a street in Salvador, Brazil.

The street, ancient and narrow, was steep and paved with large cobblestones, making walking tricky, especially backwards.

The two musicians ahead of us were surrounded by kids all playing percussion instruments strapped to their bodies. I had to keep a tight axis on the two subjects with the boom mike or their dialog would be lost in the hammering drums on either side of them, and walk sideways/backwards, as well as guide Christian down the middle of this winding fifteenth-century street.

All this, plus controlling my microphone, keeping it from violating the top of the frame as it bobbed up and down as I walked. We followed the two men for ten minutes, down that hill, until we rolled out of film just as they entered Olodum's neighborhood headquarters. Oh yes, and just before we rolled out I was able to tail slate for sync.

Soaked with sweat, high from the ride, Christian and I just looked at each other and laughed.

In the context of the film, the high levels of background noise is acceptable, for the audience knows we are surrounded by drumming kids. As long as the spoken words are crisply audible, a given level of background noise is fine.

So. Much. Fun.

THE ART OPENING: DOCUMENTARY COVERAGE vs. NARRATIVE COVERAGE

Imagine you are the sound person on a documentary about an artist.

Artists have events that show a body of work, paintings, photographs, drawings, etc., called an opening.

People are invited; they view the work, drink wine, chat, gossip, all the things that happen in a social event.

Your artist is present, and you have placed a wireless microphone on him that is sending a great audio signal back to your mixer/rig.

You have also wired the gallery owner, and the artist's boyfriend. You have three channels of wireless to listen to, mix, record singly, whatever.

You have your boom mike as well.

You and the cameraperson begin to cover the event.

- You get an exterior of the gallery and record sync sound of the street scene, cars passing, the guests arriving.

- You enter the gallery with the cameraman following the artist. As he is greeted by colleagues, you use his wireless as the primary sound source, supplementing responses to his conversation from those not miked with your boom.

- Wire on one track, boom on another. You both cover the event spending time with each subject, building small narrative portraits of them as they move about the room. This is standard operating procedure.

What else can you do?

This party is most challenging, for if we were participating in the opening, we would understand most things and probably have a pretty good time. Replace yourself with a microphone, and play back the recording, and it would be pretty unintelligible.

First, we as humans have the ability in a crazy sound arena like a party to pick out and distinguish words both spatially and socially. Spatially we focus on what someone is saying and concentrate on the conversation, using visual clues; socially, we just know how conversations work and fill in the blanks ourselves. The wireless mikes isolate conversation somewhat. Remember the omnidirectional lavaliere: things that are close are prominent, things in the distance fall away. It allows you to pick up not only the wearer of the mike, but also in most cases the person with whom they are speaking. In a documentary context, the difference in levels could be tolerated or slightly equalized in post.

The additional work you could do to make this scene work for the editor is to make a few passes through the crowd, lingering on the fringe of groups with your boom mike. Socially, people in a group situation accept the presence of a team of film makers with camera/sound, usually ignore you, and tend to be cautious about what they say whilst "on." It's cool you are filming, it's an opening, "How do I look?", and so on.

If a sound mixer walks through the crowd with a mike, it might not be so acceptable; in fact it might be considered intrusive. You are not going to boom from overhead in full press sound mode. People will feel uncomfortable, intruded upon, and quickly clam up.

You must be subtle, boom at your shoulder—after all you are just getting ambience, not the state of the union address. Make a couple of passes through the room while the cameraman has a snack. Record some tone of the room/event from a fixed spot for at least 30 seconds.

- Audio slate everything, or the editor won't know what's going on.

- At the same time you are working on your own, continue to occasionally monitor your PPL pre-faders on each of the three individuals you have wired.

What's a PPL again?

That's a switch that allows you to activate any input on your mixer to monitor what's happening on that channel. It is monitor only, so the information you hear in your headphones is not being recorded.

What if you hear something good?

Bring it up into the mix, get the cameraman's eye and go shoot the scene with the gallery owner having a fight with the artist's boyfriend, or whatever.

You just discovered a moment that can drive the narrative of the scene or the film.

How do they do scenes like this in the movies?

The beginning is pretty much as we did it. Although the exterior may be in an entirely different location.

For the main scene, depending on the director, a wide shot with the actors speaking their lines is shot as the gallery crowd mimics conversation. They move their mouths, but really say nothing.

In the close-ups, it's the same. Actors run their lines, the crowd mimics laughing, conversation, and so on. If the floor is hard and noisy, background actors visible in the close-ups may have their shoes removed or foamed.

What is foamed?

Foot foam is a thin layer of foam rubber with adhesive on one side. It is cut to fit an individual's shoe to silence their footfalls. This duty usually falls to the utility person of the sound team.

It is important to speak with the first assistant director and ask for room tone with the non-speaking crowd present and moving. This is key.

Right after you finish this, get "walla track."

What?

Walla Track is when trained extras utter noises very much like conversation. Walla-walla or babble track is easy and most extras get a kick out of doing it.

You have just gathered principal dialog, any additional off-screen dialog, quiet room tone with crowd, and walla track. Using all this, with the usual addition of a music bed, the sound editor will create a very realistic gallery opening scene. Almost as realistic as the documentary version we learned about above.

BOOMING FOR TV AND FEATURE FILMS

The majority of the booming world lives within a specific job category in the sound department.

Boom Operator

This is a person skilled in booming and many other things as well.

The term "operator" stems from those skilled in the use of the Fisher Boom. This is a mechanical device controlled by pulleys and wheels that manipulates an articulated arm at the end of which is a microphone. These devices, once common, although still in use, are sadly seen less and less on sets.

Now on dramatic narrative feature and episodic film production, the boom operator relies on handheld booms of various types.

It must be said, although it is an obvious point, that the boom operator must always wear headphones into which is fed only their boom channel. How else will they be able to judge if they are on or off axis?

Many boom operators wear cotton or soft leather gloves to silence the movement of their hands as they change position on the extended boom pole.

The mixers I work with often double boom dialog coverage. That is, when the camera is getting an isolated shot of one cast member running their dialog in a scene with two actors, *the off camera actor is boomed as well*. The timing of the actors' scene is preserved, and any intentional or inadvertent overlaps are covered. Use it or not, it allows many options in the editing room to present a very natural flow of dialog. The down side to this is that the off-screen actor is often right next to the lens for a correct eye line. This may not be the quietest place to record dialog even with the entire camera crew being as quiet as they can, as they are still creating noise in the camera's operation.

The role of the boom operator is not just holding a microphone over the actor's head to get a clear recording of the words, it is so much more . . . as will be explained by my friend and colleague, **Kira Smith**.

INTERVIEW WITH KIRA SMITH

Kira is a New York-based boom person. She has boomed six seasons of *Law & Order* and many feature films including: *Lost in Translation*, *John Wick*, *A Mighty Wind*, *Eternal Sunshine*, *Spider Man*, and *Across the Universe*, among others.

What is a boom operator?

A boom operator holds a microphone on a long pole or boom over the actor speaking just in the right place.

Why don't you just wire everybody up?

Because the boom mike sounds better, more natural.

Why did you choose to be a boom operator?

I started off as a production assistant, I was really terrible at it. Then I became a locations assistant, which wasn't for me either since it seemed like it had nothing to do with the actual shooting of the film.

I became a boom operator by lying and saying I knew how to do it when I actually didn't. That first film I worked for free so I figured they got what they paid for. I did a couple more freebies and I got better and I realized I really liked it.

The sound department is a small department but a really important one. I just kind of took to it.

I like to listen, I guess.

It's also great if you are interested in the whole process because you can't get any closer unless of course you are the director or the actor. There's a lot of times I'm the only person in the room with them . . . you know the camera is outside, the director is at the monitor, and on the set the actors are doing their work and I'm in a corner standing on a box trying to be invisible but seeing first-hand the choices the actors make, the difference that direction makes . . . you know, you've got a front row seat to the whole process, the meat of it. I like that.

What makes a good boom operator?

A good boom operator, I feel, keeps a low profile, but pays attention to everything, and can anticipate what the problems will be, and they have a good relationship with the rest of the crew, because there's no way you can be a good boom operator unless you get other people to help you because it's a team effort and we are sort of at the mercy of a lot of other departments doing their job properly . . .

So a good boom operator has to pay attention, has to know where the problems are and has to be able to come up with solutions.

For example: a camera operator is your most important relationship.

You need to know what he sees. Sometimes the only way to know that is to ask him: "Do you see this reflection? . . . Is this shadow in your frame? . . ."

You can also tell some of these things by looking at a monitor but really the best way is to get that information straight from the horse's mouth . . . so you need to have a good relationship with them.

The grips are your second most important group.

What exactly is a grip?

A grip is the guy on set who builds the thing that you need to get the shot. It's a piece of scaffolding, a dolly track, a platform, a big frame with a silk spread across it to soften the light, they are the guys that put up the big things that make everything work. They are the guys that push the dolly, that slap something together for the actor to stand on, the ones that stop things from falling on your head, they do all the heavy stuff, the moving and the shaking.

Your grips are the ones who are going to help with your shadow problems with flags, and applying black wrap creatively, and the like.

They also supply you with a ladder if you need one or with apple boxes to gain height . . . they are also the primary noisemakers on set, so if you don't have a good relationship with them, they may not work as hard to stop their own clunking and banging. They've got dolly noise, they have belts full of tools . . . you want them to be in tune with your needs as much as possible.

The physical act of booming . . . doesn't it get hard?

It can be hard, but if you have been doing it for a while you get the muscles and you find ways of not being tired. Generally the shots don't last a really, really long time, so if you find yourself in an uncomfortable position, you know it will soon be over.

What is the first thing you do when you step on to a set?

The first thing is, I have to know what the shot is, so . . . I have to know how big is the frame, what's in it what's not in it . . .

The second thing is where the actors are going to be in that frame.

Are they going to be close together? Are they going to be far apart?

Can we do this with one boom or will we have to use more than one?

Can I reach everyone?

Do we have to put a wire on someone?

That's the first assessment, and then after they start lighting it . . . then it becomes:

"Is this light going to make me cast a shadow someplace?"

"Do I have reflection issues in the windows or in any of the objects on the tables?"

Those are the things I look for: reflections and lights and what's going on in the frame.

Can I get it, or do I need help?

A bad boom operator can seriously slow down the whole train.

If I'm getting in the frame, if I'm throwing shadows, if we have to take a lot of time to accommodate me, that affects everyone.

Could you talk about the boom operator being the scout, the source of intel for the sound department?

It's generally expected that I will be the one who will let the rest of my department know what we are up against. Oftentimes the other two members of my department are doing other things like trying to kill unwanted noise, wiring up an actor, or doing other maintenance, so I am the one who generally goes to set to gather intelligence about what do we need to do, what are our problems, what do I need help with, etc.

I think other departments don't understand what we do or why we do it and it looks like we are doing a whole lot of nothing. They don't understand why we can hear some things and not other things. Why are we putting carpet on the floor? Why are we stuffing a sound blanket into that window? We are really one of the least understood departments, I think. We are so small it's easier to kind of run us over.

We don't have a big presence. We're kind of like the nerd squad. The other departments really don't know what we're doing.

It's just such a visual world, I think a lot of people don't pay attention what they're listening to.

You go into a room, there's a radiator hissing, there's a clock ticking, these are all things you block out, every day. We as the sound department go in there and we're zeroed in on that stuff. We don't want the radiator to hiss, we don't want the clock to tick, we want the soundtrack to be clean, so we're listening to what other people don't pay attention to. They don't understand why it's such a bad thing to have those noises on your soundtrack. They don't understand about the continuity of sound from one shot to the other that you can't have hissing on one side of a take and not hissing on the other. Not everybody thinks about what they hear, every day.

How do you figure out where to be in a room in order to boom the shot? Are you good at geometry?

Well, if you have seen the rehearsal and you know where the actor is going to go . . . It's veeery important to see the rehearsal . . . Sometimes you find yourself on a set where they don't rehearse, or they have a rehearsal and you're not allowed in . . . Even in this case I always find a way to stick my head in . . . If I'm ever aggressive on set, it's about seeing rehearsals. I try not to be aggressive about a lot of things, try to keep a low profile, but they really need to show me what's going to happen. So, you fight to get your rehearsal and after that you just basically have to position yourself where you can get the most dialog.

You also have to figure:

"Can I be there?" "Is there a light going over my head?"

There are a lot of factors to figure out.

I think it really helps if you play pool, for example . . . which I guess is geometry. I didn't do well in geometry [at school] . . . but I understand what you're saying.

Reflection wise, it's always a trial to figure out:

"Am I going to be in a refection here or am I not?"

General rule of thumb:

"If you can see the camera's reflection, it can see you."

Geometry and that reflection thing . . . there's just so many things that decide where you are going to be on a film set.

The most frustrating thing is you can't even go to that magical spot you want to be until the lighting has progressed. A lot of the times you have to make your mind up last minute because they light things last minute. Sometimes you have to change your plan and go to Plan B.

Like: "If I get knocked out of this spot, I can go over there . . ."

You just have to stay on your toes . . . it's sort of mentally draining to keep on top of these things.

How do you figure out where to be?

It's about reflections, it's about lighting, it's about getting the most dialog, fighting for space with the focus-puller, and being prepared if the key grip is going to come in last minute with a bounce board, plan B.

When you go into a scene you have to wait long enough to have enough information to make the right decision, but you can't wait too long because once you are in your position you have to then negotiate . . .

"Can we block this? Can we get rid of this light?"

You have to start working on making your little spot work. I think that's what takes experience is just knowing "when."

When do I put the carpets down? When do I ask the camera operator about reflections? When can I assume that it's lit, and I am not going to have to move? When do I tell them we should wire? When do we put the foot foam on the actor's shoes? There are all these things I want to do and all these people in my way! (*laughs*)

One thing that Kira didn't mention is patience. She is one of the most patient technicians I have ever worked with.

Sometimes you need to be patient.

I was working on *Law & Order*. We were in the studio on the courtroom set. I was booming Sam Waterston who is playing the district attorney. Because of the desks, chairs and railings that make up the courtroom landscape I have my boom almost fully extended, about 18 feet, with a Schoeps hypercardioid on the end.

It's a pretty tight shot, just the head and shoulders of the actor who has just one line: "Your honor. I object!"

I have worked with Waterston before and he can be unpredictable in his reads, changing voice levels, moving his head—he's kind of a wild card. With him, you have to be ready for anything.

Today, we did a series of five takes of Waterston sitting at the prosecution table saying this line without cutting camera.

I was holding the mike about six inches above his head.

On the last read he jumped straight up out of his chair like a rocket, right into the foam pop screen of my microphone.

Bam.

To say I was kind of shocked is an understatement.

Waterston turned and looks at me with astonishment, and says accusingly: "Well! We are going to need another one!"

The assistant director turned to the camera operator and asks: "Good for you?"

The cameraman nods his head, "All good."

The AD asks my sound mixer: "All good?"

The mixer said: "Good for me."

Waterston and I are both kind of astonished.

His head hitting my mike happened out of frame. No one watching through a camera, or watching the monitors in video village, saw it happen. The actor's voice was so loud saying "I object!" that his head impact was almost perfectly covered by his loud exclamation. Besides, due to the tightness of the shot, when he delivered the line, he was totally off camera, with perhaps his belt buckle or tie the only thing in frame.

They will never use a shot of his belt.

Shocked, we both looked at each other in disbelief.

But things happen quickly on set, and we just moved on to the next scene/set-up.

What happens out of frame is literally out of mind. It cannot be stressed how important it is for you to make the cameraperson aware of your boom in relation to the frame.

Often the boom operator will ask for a frame line.

What is a frame line?

That is the very top of the frame, above which the mike is out of shot.

This is when you say to the camera operator: "May I have a frame line, please."

They reply, putting their eye to the viewfinder: "Sure, put it in."

You drop your mike well into the shot, and slowly begin to lift it up out of the top of the frame.

They respond: "You're in . . . you're in . . . you're in . . . that's good, you're out!"

You look across the set to the wall behind the boom mike and mentally line it up with a picture frame, a piece of molding, anything to give you a visible reference to the "safe" area that your mike can live in.

To give your microphone added visibility a thin strip of paper tape is often added to the very tip to make it more visible.

Remember you are relying on someone—camera, script, director—to notice if your mike dips into frame. It happens. If it's dark, and most windscreens are made of black or charcoal foam, a slight dip in of the boom may not be spotted until later.

One of the most extreme examples of a cameraman "not noticing" was during a feature I was booming in Boston.

It was a lovely fall day. We were outside a townhouse shooting a character leaving for work played by the great actor Karl Malden. His wife was to say goodbye to him from the bay window as he walked down the front steps.

I was on a 12-step ladder, with the longest boom pole we had, I think a 22-footer. On the end of the pole was a Sennheiser MKH-816, a really long microphone, which was inside a plastic zeppelin to protect it from the wind.

The sun moved, as it does, and now the shadow of my pole was visible just above the townhouse window in the center of the frame. With few options available, we decided to attach some leafy branches to the pole so the shadow of my boom just looked like a tree limb. Clever. We were so proud of ourselves.

Two days later in dailies (where the rough takes with sound are screened for appraisal), the shots of the townhouse with our clever tree branch displayed our leafy shadow, but they also showed the whole boom, microphone, the taped-on branches and leaves, fully in frame, just flying through the top of the shot.

For like six takes.

After watching the second take, people in the screening room started to giggle; after the fifth take, the room was just filled with gales of laughter. The only people not laughing were the producer, my mixer, the cameraman, and me. The cameraman had changed lenses at the last minute, making it a much wider shot, and then never noticed the flying tree branch/boom pole/zeppelin rig right in the middle of his shot. We had to re-shoot the whole scene. If we had been using video assist, an external video monitor, for the director and script to analyze the scene, this would have been caught.

Production felt the quickie exterior didn't need video, so it was moved on to another location. The quickie exterior wasn't so quick any more, for we had to shoot it all over again.

The producer called us into a small huddle after the dailies. He didn't have to say much; we all were guilty as charged.

"No more boom shadows!"

Always communicate with the camera department, no matter what the shot may be.

POINTS TO REMEMBER

- The difference between documentary and narrative booming

- The importance of situational awareness: keep your head in the game!

- What makes a good boom operator

- The grip's role

- The importance of watching rehearsals

- How to get a frame line

- The importance of communication

EXERCISE

Take a small flashlight, and place it in a mike mount at the end of a boom pole.

Enlist three friends:

• one to hold a smartphone that shoots video

• two to walk out of a room and down a hallway, having a conversation.

Have the smartphone/cameraperson walk backward shooting your friends speaking as they go from the room and down the hall.

Meanwhile, you boom the two speakers.

Roll to record on the smartphone.

Action.

Pan the flashlight from person to person as they speak.

Review the smartphone footage:

• Did you stay out of shot?

• Did you keep the flashlight's circle of light on their mouths/chin/chest as they spoke?

• Did you clear the doorframe with your light/boom mike?

• Repeat.

• Repeat faster.

• Repeat slower.

How did you do?

7 WIRELESS MICROPHONES:
What They Are, How They Work, and How and Why We Use Them

They are a sonic compromise, a risk, or the answer to a producer's prayer. Wireless microphones, or radio mikes, are all these things and more. A fascinating, evolving technology that is a major part of film for the moving image is explained, shortcomings noted, and fully revealed in the chapter that follows.

Wireless microphones are an essential part of film production. They allow a great deal of freedom in certain productions.

Their unique virtues and vices will be covered at length.

WHY ARE WIRELESS MICROPHONES, OR RADIO MICS USED IN FILM AND TV PRODUCTION?

- The main reason is the frame. A shot can be super wide and dialog can still take place and be recorded without an overhead boom mike. The boom microphone, or its shadow, or its operator, is literally out of the picture.

- Many subjects can be heard clearly at once in a particular shot. You can have as many wired actors as you have tracks to record them running around in frame and hear them all. This technique was pioneered by the genius American feature film director Robert Altman.

- In reporting news and sport, the correspondent can hold a handheld radio microphone and not have to worry about a cable link to either the camera or sound mixer. In crowded venues, locker rooms, press openings, and other busy events, this is a real asset.

So it is the first shooting day of *Project Runway*, a show about fashion design. There are 16 people all wearing microphones and body pack transmitters, plus the two hosts of the show, who are also wearing wireless mikes. That's a lot of radio frequency signal in a place, New York City, that already is pretty busy with electromagnetic energy.

So, if you have 18 people, and 18 transmitters, you also must have 18 carefully chosen frequencies.

Each sound mixer who is paired with a camera has four receivers hooked up to his mixer.

This is the first day, however, and so six of the sound mixers, who are each assigned to a cameraperson, must consult their face cards, which is a row of photographs, a sort of mug-shot-line-up

of designers, identify the designer, then tune a receiver to the frequency channel written above the photo. If you have 18 people, and 18 transmitters, you also must have 18 different frequencies.

It's like each person wearing a transmitter is a tiny radio station, but instead of music and news they broadcast every word they say. It is the job of the mixer to tune that station in so those words can be recorded.

That's complicated! Can't you just record all 18 cast members on a single recorder?

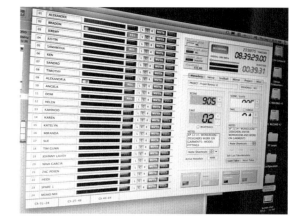

METACORDER MULTITRACK RECORDING PROGRAM

Good idea! Yes you can. Budget permitting, that usually takes place as well. However, the multitrack recording is considered a backup to the recording done directly to camera. Your job as a mixer to is to record the voices of the people your camera is seeing.

Typically, on a show like *Project Runway*, each camera has a designated sound person.

That sound person is carrying a mixer with receivers.

(In the case of the photo below, a five-input mixer with four receivers and a boom.)

Their mixer then sends a right and left output mix to his camera via wireless transmitters.

These are called *hops*.

A multitrack recorder records all the tracks, all the time, if the person is in the shot or not. They could be in another room, or even building, and if they are still in range of the master recorder's wireless receiver they are being recorded.

It takes a considerable amount of time and effort on the part of the editor to locate the track, transfer it and sync it to the picture. In TV Land, time and effort cost money.

Most of the time the sound is pulled right from the camera's disk.

Sound Devices 552 Mixer with four Sennheiser wireless receivers and a nine-person face card

The first four channels are taken up with wireless receivers; the fifth channel of the five-input mixer is for the boom microphone. This is usually a shotgun/supercardioid/hypercardioid on a boom pole.

The boom mikes on this kind of production are usually Sennheiser MKH-60 or 416p.

The number of wireless receivers is limited by the number of inputs you have to work with on your mixer. In the case of the 552 sound mixer (electronic equipment), a sound mixer (human technician) has five inputs to employ in the recording of audio for video.

So you as a location mixer have five channels to mix as you gather the many voices of those who are cast members.

If one of the designers speaks, the sound person must tune one of the receivers to the frequency of that person's transmitter.

An MKH-60 with a protective foam windscreen

A long six-digit frequency is given a smaller number: say, 646.000 becomes Channel 15.

So each cast member and host has a number from 01 to 18.

Designer Monica begins to speak. Monica's transmitter frequency channel is number 07.

The sound person pushes the little buttons on the receiver until it reads 07.

OK, but there are 15 cast members all wearing wireless. How am I supposed to record 15 people with only five inputs/six receivers?

It is your job to mix only the people who are seen in your cameraperson's frame.

You must identify the person speaking by comparing them to the image on the face card, and then tune a receiver to their specific frequency.

Then you turn up the level on that receiver's input and, through the magic of radio, you hear their voice.

That voice is then transmitted to the camera via another pair of designated wireless transmitter/ receivers, and recorded by the camera on one of two tracks, either left or right.

Everything you mix is sent to the camera by the wireless *hops*.

I assume they are called hops because the sound "hops" from you to the camera.

More about hops in a moment.

Wait, I'm supposed to record sound, identify cast members, dial in their frequencies, mix the various voices according to what the camera sees, and stay out of the shot of all cameras working the scene and fly the boom at the same time. You're kidding, right?

No. That is your job. At first, it seems insane what you have to do just to get audio. After a few hours of this craziness, you adapt. *You get the hang of it.*

Why is the boom mike important if everyone is wearing a microphone?

In group situations, or if someone runs into frame suddenly, and you do not have their frequency tuned to a receiver, that's when the boom is essential. Get the sound. Stick your mike up there, boom the new person, and *then* dial them in.

Cameraman Brian Stimele and Sound Mixer Wyatt Tuzo backstage at Fashion Week

You must get the sound.

Let's go back to our first shooting day. The designers arrive at the small garden in front of the School of Design where the action takes place and begin to greet each other.

As a mixer, you may not have the correct frequency dialed in for the person your cameraperson has decided to follow. Things are happening fast, and the designers are talking a mile a minute. There are seven crews. Camera people are bouncing from person to person looking for the best story/shot. If you take time to enter the frequency, you are going to miss what the designer is saying. If you look down to dial in a frequency, you will miss your camera moving quickly to another more interesting designer, or worse, you yourself might end up in the shot.

Get that boom up there and get the sound.

With the boom up in one hand, positioned over the pair of designers having a conversation, you can record their conversation while you tune your receivers with the other hand. You have to be quick and error-free. One push too many on the tuning button and you'll miss the frequency; one moment of inattention and your boom drifts off axis, or worse dips into the frame.

Concentrate.

In some situations—say, if four designers are speaking at once in a small group—it usually sounds better to boom the group as a whole than to have four microphones open within a few feet of each other.

The most important thing is to get the sound.

You have to get every word that is spoken.

You must know how wide the shot is to stay out of the frame.

You must know where other cameras are to stay out of their shots.

A sound person has to be alert, keep focused on the work and have great *situational awareness*.

In addition to the sound person dialing in each designer speaking and booming the shot, the mixer you are using is feeding the audio, two tracks of your mix, to the camera. The camera is recording this on disk or memory card.

There are two ways to feed audio to a camera:

1. Either by a cable attached directly to the camera, called a multi pin, or breakout cable.

2. Or via two wireless transmitters sending your audio mix to a pair of receivers on the back of the camera.

These receivers are plugged into the camera's inputs one and two.

This audio being sent to the cameras is often what production uses for the final product due to time and cost reasons.

This is time to reintroduce the *hops*.

The right and left output of your mixer each has a wireless transmitter sending one channel of your audio mix to a receiver turned to the same frequency hanging on the back of the camera.

This receiver's output is plugged into one of two inputs on the back of the camera.

The camera receives the audio from you mixer of whatever your two-person team is recording.

Pictured is the basic Reality Camera with a pair of Lecrosonics 411 Hops receivers on the back. The 411's antennas are poking out of a nylon saddlebag holder for both the right and left receiver.

The cameraman is wearing a surveillance-type headset for a walkie-talkie from which he receives directions from the supervising producer or director.

99

Lectrosonic UM400a right channel
hops transmitter

Cameraman Gene Bradford follows the action

Why introduce another set of wireless when you already have like a zillion working already?

The only alternative to hops is to be tethered to the camera by an audio snake. This is a cable that combines two full balanced audio lines as well as a return monitor line. They come in various lengths. A fuller description of an audio snake appears in the next chapter.

The operative word here is tethered. A horse is tethered to a fence. A sailboat is tethered to a dock. A monkey is tethered to an organ-grinder.

In other words, you are physically tied to the camera and cameraperson. In a busy, fluid, shooting situation, this can be a nightmare.

Advantages of wireless hops

Freedom of movement for both the cameraperson and you the sound mixer. It's huge not having to worry about keeping up with your dance partner, the cameraperson.

It also allows those in post-production to quickly log and edit the footage without going through step of syncing sound and image. More about this in Chapter 11.

Disadvantages

The mixer cannot monitor what is being recorded by the camera. The camera, whether we like it or not, is an audio recording device. Once audio is sent by hops it becomes impossible for a sound person to monitor the camera's ability to record sound. You have to depend on the cameraperson to listen via a small headphone, or the small speaker built into the camera, to ensure the quality of the tracks. Although we are a team, this is not really their job. They have to shoot, focus, and listen to the producers on the walkie channel for guidance regarding their next shot.

In some cases, an additional wireless sends the return headphone output to the mixer so he can monitor the camera.

Another wireless?

Wireless systems are not immune to interference. You might get hits or dropouts in the hops system. You might get dropouts in the monitoring wireless. Who's to know where the hits are coming from? Also an additional wireless clogging up the already jammed airwaves may not really be worth it.

Actually, if you have tuned your hops to a clean frequency band (more on this below), and you are standing only a few feet from the camera and your receivers, things are usually pretty bulletproof. Keep an eye on the levels visible on the side panel of the camera to make sure both channels are working, and check playback. *Often.*

That's pretty intense!

Yes, that is one busy group of sound people. As the show goes on, you get to know the faces of the designers and their frequencies on the card so well that you do not need to look at the card as much. Also the designers begin to be eliminated (there are fewer frequencies to tune!), so it's not as much of a scramble as it is at the beginning.

That's a lot of wireless transmitters at work in one place.

OK, how do these wireless things work?

Again, it's all about waves.

A transmitter combines a voice from a microphone with a radio signal into a special combination radio signal which is then fed to an antenna.

These radio waves are modulated. This means that the original audio signal is superimposed on the radio wave so that the wave "carries" the audio.

This electrical signal makes the electrons in the metal of the antenna change energy levels and emit radio waves.

A receiver is really a transmitter in reverse. Radio waves strike the antenna connected to the receiver. These waves affect the metal atoms in the antenna which reproduce the carrier signal from the transmitter. The receiver then selects the carrier signal of the channel. It then extracts the sound signal from the carrier signal and sends it to an amplifier to reproduce the sound.

As we learned in Chapter 1, all waves have a particular wavelength or frequency.

Frequency, of course, is the number of waves that are transmitted over one second, measured in Hertz (Hz).

Radio frequency radiation is measured in Kilohertz (kHz), Megahertz (MHz), and Gigahertz (GHz).

The frequency of the carrier we use the most exists in the **VHF** and **UHF** bands.

- VHF Very High Frequency: 30 to 300 MHz
- UHF Ultra High Frequency: 300 MHZ to 3 GHz (3000 Mhz).

UHF systems do not hold any large technical advantage over similar VHF systems.

The big advantage of UHF is that there is less chance of interference because there are fewer transmitters operating at frequencies likely to cause problems.

When two transmitters operate at the same frequency, you experience *intermodulation*. This is known as intermod, or IMD (intermodulation distortion). It sounds like a mistake. A big confused electronic mumble.

Electrical interference due to other devices is also generally lower at UHF frequencies.

This is because noise from these sources becomes less intense as the frequency increases.

Interference of all types does not travel over as great a distance as at VHF frequencies.

VHF systems are usually lower in cost, as well as consuming less power (less batteries).

Use of VHF in a dense urban environment is considered somewhat risky (and frustrating).

Both VHF and UHF transmit best when a clear path is visible from the transmitter to the receiver.

This is known as line of sight transmission. Walls, buildings, chain link fences, hills and other physical barriers can all cause signal degradation. Airborne moisture such as rain can affect transmission as well.

Lectrosonics plug-on transmitter plugged into an Electrovoice RE50B microphone

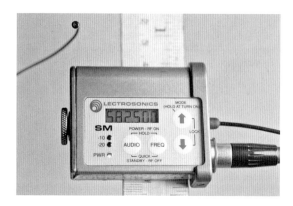

Lectrosonics SM miniature body pack transmitter

WIRELESS MICROPHONES—THEY'RE GREAT! UNTIL THEY STOP WORKING

What do you mean, stop working?

The first and most annoying event that occurs is when you are rolling on a scene and the wireless transmitter/receiver begins to be affected by interference.

What does that sound like?

Static pops hisses, or complete signal dropout that can make your day miserable.

How can I avoid this, and if it happens, what do I do?

The first thing is to be aware of where you are. If you are on a wheat farm in Nebraska, chances are that you are not going to have to worry about interference from other frequencies or physical barriers between you and your subject. Urban America? Watch out.

Before you start your shoot, do a frequency scan.

If your wireless has the ability to scan the airwaves for competing frequencies this can be a very useful tool. The receiver actually reads electromagnetic currents in your area and by using a bar graph, tells you where are the best places to tune your receiver and transmitter.

This resource can be very helpful, for in the metropolitan New York area, reception can vary block by block, or even floor to floor within a building.

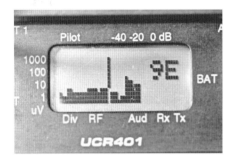

Lectrosonics UCR401 frequency scan level 1

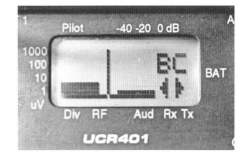

Frequency scan level 2

The Lectrosonics system allows you to scan on two levels.

The first gives you a graphic picture of what parts of the channel are in use.

The second allows you to narrow down your search for a hole, an unused, or less used part of the frequency spectrum. Once you find a valley in the graphic display, stop the scan, hit all three panel keys, and in answer to the display prompt "use new."

Be sure to tune your transmitter to the same frequency as your receiver!

Double check that the numbers/letters are the same on each.

What else makes a transmitter stop working?

A sound person has to control their world. But you really must control *two* worlds:

1. recording sound

2. systems maintenance.

Recording sound: The first is obvious. You have to be able to make the right choices and record awesome tracks.

Systems maintenance: This is just as important. Without being good at the care and feeding of your awesome tools, it might mean that you will never be able to record those fabulous tracks.

With wireless transmitters, it is always *batteries, batteries, batteries.*

At the start of each day, put in fresh batteries.

At lunch, put in fresh batteries.

Turn off the transmitter when there is going to be a lengthy break, such as travel or a tedious relight.

A good sound person always carries replacement cells. Things do not always go on schedule. Most wireless receivers have a display that allow you to monitor the transmitter battery voltage at the receiver.

If the battery starts to get low, replace it.

There is nothing worse than being in the middle of a scene or interview and getting a flashing low battery display.

Unless you want to interrupt the scene, you are forced to wait for a break to refresh the battery in the transmitter. That's when a break occurs—when a disk or media card is switched, or when the interviewer pauses to ask a question. If you don't act, you risk losing your track, or all of your audio when the transmitter finally dies.

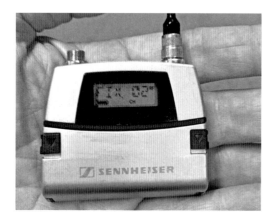 

Sennheiser SK5212 Wireless Transmitter

Many body pack transmitters have variable, selectable power output.

What does that mean?

You can control the power level of the radio waves from your subject.

Why is this important?

The more powerful the transmission the less likely you will be to get dropout or a hit.

Why don't we just run full power all the time?

The situation may not call for full power. When you run at full power, it's like driving a car at full speed.

You go fast, but the tank gets empty really fast as well.

For the transmitters we use in film production, the transmission audio power is measured in milliwatts (mW).

Usually the three output power choices are:

• 50 mW

• 100 mW

• 250 mW.

50mW is great for theatrical productions or interviews where the receiver and its antennas are nearby.

Like on a stage or studio.

100 and 250mW is used for extended range when the talent is far away or the shoot is in a very active electromagnetic area.

250mW is the most powerful. It blows away weaker interfering signals and allows greater range between the subject wearing the transmitter and the receiver.

However, when transmitting with more power, you use more power.

That means your batteries drain quickly at higher output settings.

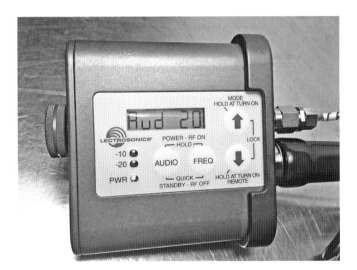

Lectrosonics SMQV Transmitter

These are the battery power consumption numbers below for the Lectrosonics SMQV:

		Alkaline AA	Lithium AA
SMQV (two AA cells)	Transmitting at 50mW	6 hours	14.5 hours
	Transmitting at 100mW	5.5 hours	14 hours
	Transmitting at 250mW	1.7 hours	7.5 hours

So you see, running at 250mW really sucks the current out of your batteries. Life is good when you aren't worrying about batteries dying.

The chart also shows you how much better lithium cells perform than alkaline AA cells.

Lithium cells

Sure they cost more money, but are you going to risk not being hired again because of always stopping for battery changes or dead transmitters just to save a few dollars? *Really?*

How to save power extend battery life: turn off the transmitter!

Some wireless transmitters can be manipulated (put to sleep, change audio level, or even change frequencies) by audio tones alone. This works by holding an audio tone generator near the microphone worn by the talent and sending a sound, sounding very much like a fax signal from a phone.

There's an app for this . . . It's called LectroRM.

You can even be a few feet away, press a button, the generator chirps, and the transmitter goes to sleep, cutting battery use to next to nothing. The same app will change frequencies, audio levels, and lock the functions without you ever touching the transmitter.

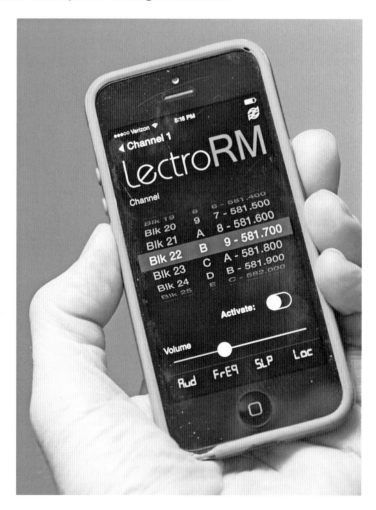

LectroRM App

Across the bottom:

Aud: Allows you to remotely adjust the sensitivity of the transmitter.

Freq: Allows you to remotely change frequencies.

SLP: Remotely puts the transmitter to sleep.

Loc: Remotely locks or disables any of the controls on the transmitter.

This is an awesome thing if the transmitter is buried in a costume, or in a difficult place (like a thigh or ankle strap) that would require the talent to partially undress for you to access the controls.

The less you disturb an actor, the better.

On a recent *The Good Wife* shoot, I was able to "sleep" the wireless of Chris Noth and Alan Cumming as they sat and studied their lines just by standing a few feet away and chirping the audio tone into their microphones. The noise attracted Alan's attention.

I explained:

"I was just putting you to sleep."

He then pretended to fall asleep.

Actors do appreciate the respect for their privacy on set.

Trouble!
OK, my batteries are fresh, but I keep getting hits on one of my wireless!

The first thing to do is move! Like the title of the opening chapter, without movement there is no sound, the first thing to do is get your receiver closer to the transmitter. Even moving a couple of feet can increase reception or block the signal of the interfering electromagnetic energy.

I was working on a BBC documentary in the late 1980s about the drug war in Upper Manhattan/ Washington Heights.

We were in an unmarked car on a stakeout with an FBI agent who was monitoring a police radio frequency. His incoming police radio signal kept breaking up to the point of no longer being understandable. The detective put the car in gear and crept forward two feet. The signal was better. He then put the car into reverse and crept back about three feet. The signal was perfect. This is something he dealt with every day. *Secrets of the pros!* Just a matter of inches can solve your problem. Or save someone's life.

Do something. Anything. Turn around. Touch your antennas with your hands. Move forward, or backward . . .

If the interference occurs during a key take, *always tell the producer/director about it.*

SPEAK UP IF YOU HAVE A PROBLEM
If you get a wireless hit, or have an audio issue of any sort, it is most important to notify production at the end of the take.

You may be the only one to know or notice.

Sound is your medium, so it is your call.

Even if production, or the camera operator is monitoring with IFB, Comtek, or Icom listening devices, never take for granted they hear the same thing you do.

This is key. You are being responsible for your domain.

If you do not speak up, and just hide under your headphones, you will never be taken seriously as a sound person again.

This is not to say if you hear a hit, lavaliere rustle, or fire truck, you are to jump up and wave your hands, calling a halt to the proceedings.

If it was in the middle of a sensitive documentary interview where interruption would burst the bubble of a moody or emotional narrative, this would be a judgement call. You'll know if, say, the subject is in tears, and the audio has an issue, that you just may need to let it go.

It's your call.

Speak to the director afterward about the issue.

Similarly, if you hear an oncoming ambulance that is soon to fill the audio sound scape with crazy sirens, silently get the producer/director's attention, point at your headphones and shake your head. Let them call it. You did your job by calling attention to the issue.

If you are recording multiple takes on a commercial or narrative work, call out after the flawed take: "We need another one."

We need another one. The "We" is important.

Camera gets another take when they blow a shot; you should too.

Be persistent. If you have a problem with lavaliere rustle, and you attempt to fix it numerous times without success, keep at it until you work it out.

No one can fault you for being professional.

Then again, sometimes you just have to live with it.

I was working for a 60 Minutes story about Mob guys/organized crime. We were in a very mobbed-up apartment in New Jersey. I had two of the subjects wired up. Everything was fine, except every five minutes or so, I would hear an electrical snap in my headphones. I tried everything: switching the lavalieres, the transmitters, different receivers, new frequencies, all without success. The answer was right in front of me, literally. The mob guy apartment was carpeted in a long, brown polyester shag. The mob guys themselves, although clearly not athletic, were both wearing nylon track suits.

As they would walk around on the carpet, the static electricity would build up in their nylon traxedos until it discharged along the wire of my lavaliere. What to do? Clearly the carpet had to stay. We had already shot for an hour with them wearing the nylon warm-up suits. I mentioned it quietly to the producer. He just smiled, said OK, and we continued.

WHAT IS AN A2?

An A2 is a job classification born mostly as the result of reality TV. An A2 has a number of duties.

Primarily his job is the care and feeding of wireless microphones.

A good A2 is a specialist at wiring talent. That is, placing the microphone, the transmitter pack, and it's connecting wire on a person's body so no one can see it.

We will talk more about the A2 and their duties at the end of this chapter.

I first became aware of the A2 position while working in the early days of reality television. There was a BBC reality show about staging a Broadway musical production. I was called in by the sound supervisor to wire people in a small room at the foot of a Broadway stage just before they auditioned to be on the show.

The tryouts were scheduled for two days. There were approximately 150 young men and women who were to get up, one by one, on an empty stage, and sing. Each audition took about five minutes start to finish.

The mikes had to be hidden, never to be seen on camera. The offstage room was large enough to hold five promising singers, myself, and my assistant.

We had six wireless mikes.

It was like an assembly line. As the person would stand in front of me, I would assess what they were wearing and place the TRAM microphone (1) in the appropriate place to sound the best, and (2) in the best position to avoid clothing rustle.

By the time the singer was done, I was finishing with the next person and my assistant was removing the mike from the one who had just auditioned. Clockwork.

I had to make the right decision for placement, and be fast in positioning the lavaliere. Two to three minutes tops for each person. Over two days and 250-plus people, I got very skilled in wiring people in a very short time. Practice . . . practice.

Being an A2, and wiring, is kind of an art. You have a selection of tools and materials to work with, a choice of microphones, and tricks for disguising transmitters and their antennas.

Creativity is very important

Wiring the talent is just one of the duties of an A2.

On some shows, the A2s are also systems installers and troubleshooters who handle an array of tasks, they are really a production sound electrical engineer.

The heavy use of wireless microphones necessitates the placement of antennas throughout the location where shooting is to take place. Heavy antenna cable has to be run, sometimes through ceilings. The multitrack/mixing console suite has to be uncrated, assembled, and wired correctly. The 15 or so individual sound persons' mixer rigs have to be assembled and tested. Booms must be assembled and tested. Individual interview rooms must be soundproofed and prepared for shooting.

All wireless transmitters and receivers for both the talent and links from mixer bag to camera must have the correct frequencies programmed and tested.

In-ear receivers must be tested so the director can cue then camera talent remotely.

Portable listening devices (IFBs) must be programmed for the correct cast transmitter frequencies as well as the proper hops transmitter frequencies for each camera crew.

Stocks of vital elements need to be organized: expendables: batteries, tape, A2 supplies, replacement cables, and spare parts, all logically arranged for quick access.

Being an A2 means paying attention to a lot of details.

When shooting starts, the designers, who stay in separate hotel suites, are wired by an A2 first thing in the morning, and un-wired just before they go to bed.

The show I work on is about fashion, and it is common to be challenged by the style/design of clothing that is worn by and designed by the designers.

In other words, what they choose to wear is often a daily pain to deal with.

Jewelry, glasses, measuring tape, strips of material are all hung around the neck.

The judges of the show often wear outrageous haute couture dresses and jewelry which makes microphone camouflage and placement very challenging.

A sound person or A2 has to, by the nature of the job, become rather intimate in placing the microphone on the judge or talent. You literally have to do you work under someone else's clothes. Sometimes a TRAM lavaliere with a vampire clip stuck on a square of molefoam has to be placed on the exact center of the bra under a low-cut blouse or dress. Right in a woman's cleavage, right down Broadway. In other instances, the only way a transmitter can be hidden under a tight-fitting dress is to use an elastic thigh-band that has a pocket sewn into it. It's personal.

Be professional. It is an awkward situation for both yourself and the person who is being wired. Be confident, quick, and always courteous. Assure them that if the transmitter or lavaliere causes any discomfort, they should have production send word to you, the A2, and you will come to their aid. You are there for them. On most productions, the A2 does the majority of wiring and adjusting of the microphones. If one of the sound people notices that a mike has shifted, come loose, is being rubbed by an added scarf, or a mike cable is peeking out from under a shirt, you are the one to correct the situation.

When wiring children, it is always good to have a parent nearby: it calms the child and protects the sound person from undue scrutiny involving the adult–child interface.

Vampires, moles, and toupees: the fine art of wiring people
The most common wiring jobs you see, literally, is when a microphone is placed on the outside of the on-camera person's clothing.

News anchors, correspondents, talk show hosts . . . There is no illusion to be maintained, as these people are clearly in front of a camera speaking to the audience, so the lavaliere is visible.

Simple, visible, the lavaliere is right out in front, in the clear.

 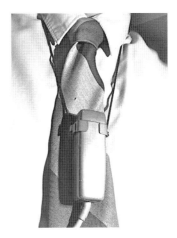

The pioneers of the lavaliere world: Sennheiser MD 214 and the Sony ECM-50

With a tie clip, there are only a few things you have to consider as potential problems.

Security is the first.

Having the mike wire show from behind the tie looks messy and unprofessional. Always secure the wire with some tape.

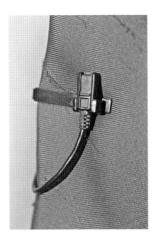 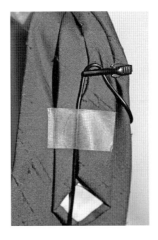

The most useful kind of tape is the plastic surgical tape for medical/EMS use. The brand I use is 3M Transpore. It and other generic brands are readily available in drug stores. The stuff is incredible. It sticks forever, to anything, especially fabric. I have had a piece stuck on my shirt survive a full washer and dryer cycle and still hold tight.

There are other types of first aid tape as well, foam-based, flesh-colored, etc. Be creative!

Once you have positioned the lavaliere just the way you want it (not too high, not too low), the other potential problem comes from the actual talent themselves.

People speak in many different ways. It might be that the person you have just wired has an explosive way of speaking. That is they puff a lot of air out during the pronunciation of certain words. This can cause your little lavaliere to overload with a blast of breath.

Then it's time for a microphone wind/pop screen. Constructed of foam or wire/foam they are built to shield the mike diaphragm from just such issues.

They are not much help in real wind outside, however. We will address serious wind screening later in this chapter.

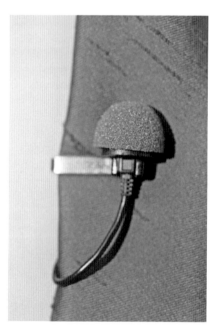

If the talent is wearing a sport coat or jacket over the shirt and tie combo, a sound person can also position the lavaliere on a lapel. The one thing to consider in this situation is which side to place it on.

If the mike is on the left lapel, and the talent's co-anchor is on the right, every time the talent turns to their partner while speaking, there will be a drop in volume. If two people are on camera, determine in advance which side the talent will be turning and position the lavaliere accordingly.

Generally, I try to avoid this kind of asymmetrical miking.

It's time to go into hiding . . .
Certain shows and fiction productions demand the illusion of life without everyone bristling with wires and windscreens.

Your life just got more challenging.

Let's talk about clothing and camouflage.

In some situations, say, a sit-down interview, you only have to disguise the lavaliere and its wire. Everything else, like the body pack transmitter, is out of shot because the shot stops at mid-chest. In other situations you have to hide the lavaliere, wire, and transmitter body pack, all day, from cameras shooting the subject from all angles. This also means your initial wiring job has to stand up to constant work, movement, perspiration, bathroom breaks, all the human activity that a normal person being followed by four camera crews would go through in a 14–16 hour period.

Let's start simply, shall we?

The basic shirt/blouse hide

Let's start with a nice cotton shirt. No tie. Hide the lavaliere.

My go-to microphone in this situation is a flat-mounted side address mike like the TRAM TR-50. Other manufacturers make similar products, but the general design is important here, not the make.

First mount the *vampire clip* on a piece of *moleskin*.

What? Vampire?

Yes, *vampire*. The vampire clip is a plastic holder that on the front grips the sides of the microphone and with two pins on the back face, is designed to stick into cloth. This is a brilliant design that we are making even more secure by inserting the vampire's pins into a piece of moleskin. Moleskin is something you usually find in the foot care section of the pharmacy. Sticky on one side, cloth-like on the other, its usual task is to isolate blisters, or take the edge off the back of a boot . . . something foot-related. We buy them in sheets and cut them into little lavaliere-sized squares. There are two types, the regular moleskin and molefoam. Molefoam has a thicker foam backing. Both are invaluable. For even more camouflage, you can obtain the rare black moleskin from audio supply/sound rental houses.

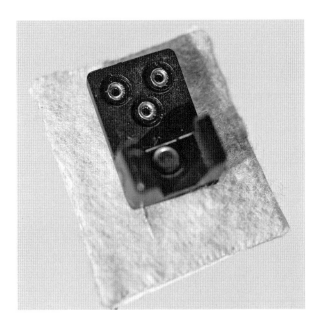

Right, so we have mounted the clip onto the moleskin square. Now insert the TRAM lavaliere into the clip with the grill/diaphragm facing into the clip itself. Although it may not seem so, the small space between the microphone and the clip is enough to allow the mike to work properly. Additionally, the back of the TRAM, which is solid plastic, provides protection from clothing noise.

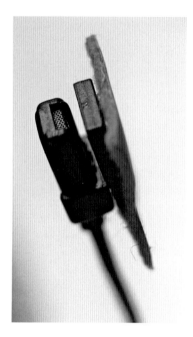

Now place a small strip of Transpore, just the width of the mix clip, to secure the lavaliere to the clip.

If the subject is not moving, say, sitting in a chair, you should not always use moleskin, but it's just become a habit with me.

Now take the TAF5 (F) connector on the end of the lavaliere and drop it down behind the shirt underneath the buttons/placket.

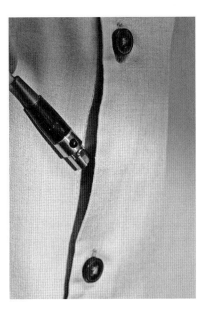

You have to be quick as you wire the talent. Wiring is not just placing and securing the microphone, but also feeding the wire and connector through layers of clothing to the transmitter as discreetly as possible. An aid in doing this is adding a weight to the TAF5 (F) connector so it drops with more authority between layers of clothing, pant legs, pant waistbands, etc.

I have machined my own TAF weight; however, a similar product is available called LavBullitt with adaptors to fit all popular connectors.

Firmly stick the moleskin patch to the inside of the outer shirt placket. Other than a piece of Transpore on the wire somewhere downstream from the lavaliere for strain relief, you're done with the microphone end.

Now, for the transmitter . . .

If it's a casual interview, the sound person can just run the end of the microphone wire through belt loops, or the skirt waist and then clip the transmitter to the belt or inside a pocket.

On women, a favorite place is to clip is to the back strap of the bra. This is a favorite for a number of reasons:

If you need to change a battery, it is relatively easy to access. Also, mounted on the top half of the body, it's not in danger of damage during a bathroom break. Transmitters have been destroyed by being dropped into water.

Yes, *that* water.

If someone is wearing a transmitter all day with a lot of activity involved, then you use a neoprene belt with a pocket built in, or an elastic strap with a cloth pocket. These are both very secure; however, keep in mind that they are wide belts that encircle the talent's waist, ankle, or thigh, so for everyone's sake they must be laundered regularly, to cleanse them of hours of performance anxiety.

Top to bottom

- Ankle pouch

- Waist pouch, Lectro SM transmitter, Lectro UM 400 transmitter

- Ankle strap

- Waist strap

The order of wiring while using a belt is as follows:

1. Mount the microphone.

2. Turn on the transmitter.

3. Insert the transmitter into the strap pocket.

4. Wrap the belt around the talent with the transmitter facing inward toward their body.

5. Plug the lavaliere cable into the transmitter.

6. Carry out the sound check.

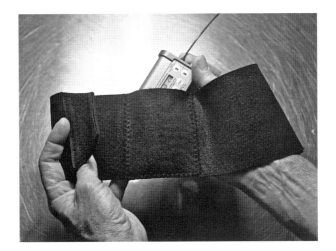

Tie one on!

If you must do a considerable about of concealing lavalieres, then you must be like a fly fisherman and have a box of different microphones for specific clothing challenges.

Some lavalieres can handle being under a light layer of clothing, and some really don't work well at all unless completely exposed.

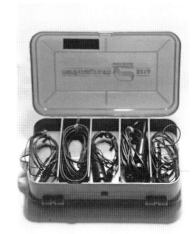

TRAMs are excellent lavalieres. All the mikes I own are excellent. However, the design of a TRAM would really not be the mike for our next gag: Hiding a lavaliere in a tie requires a design that is physically different than a TRAM, which is a *side address* microphone. This refers to the sound being "addressed" by the microphone from the side versus from the front/top.

This is not the ideal application for every situation, but then there is no-one-size-fits-all microphone. That's why a sound person should carry a whole bag full of solutions.

Left to right

- Countryman B6   Top address
- Sanken COS-11   Top address
- TRAM TR-50   Side address
- Lectrosonics M152   Side address

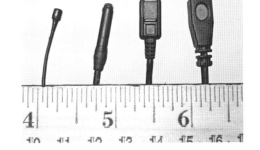

My go-to mikes for tie rigs are the Sanken COS-11 or the Countryman B6. They are both top address microphones.

The Sanken COS-11

The COS has a long rigid body. At the end is a small wire cage under which is the diaphragm.

It is an excellent sounding mike if it has plenty of air around it.

Its length and rigidness make it easy to push through a tie knot.

To prep the COS for a tie, take a small piece of black moleskin and wrap it around the mike just below the wire cage.

Feed the microphone behind and over the tie knot. Then insert the COS just behind the front fold of material until it emerges beneath the knot. *Now things get sticky.*

There is a product invented to keep toupees and wigs on the wearer's head. It is called Topstick. It is basically clear, rubbery, double-sided tape that is really sticky.

We sound people have adapted it, like Transpore, for our own needs.

Take a strip of Topstick, tear it in half, fold it over, then wrap it around the moleskinned COS-11.

This prevents the COS from slipping back into the tie knot as well as further insulating it from material rustle. With this combination, that mike is not going anywhere.

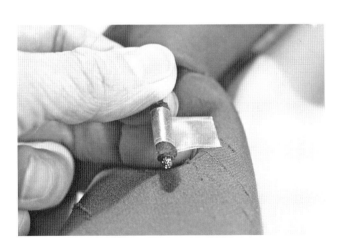

Small but mighty . . .

In some situations the answer is not to bury a mike but to have it live out in the open. A sound person can get away with this only if the microphone is so tiny that it becomes invisible against a background of clothing, hair, jewelry, etc.

It is actually hiding in plain sight. The mike I use for this is the Countryman B6.

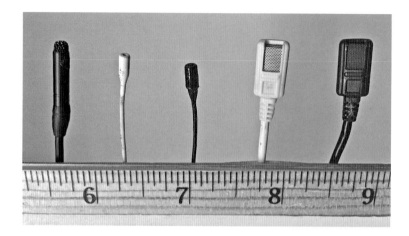

You can see how small the two white and black B6 (second and third from the left) are compared to the COS 11 and TRAMs.

Although the B6 comes provided with a tie clip, the sound person really has to improvise to mount them successfully.

The size and versatility of his mike has saved me countless times from costume/fashion challenges.

On one occasion, a judge on a fashion show was wearing an enormous square necklace/pendant mosaic around her neck that was the size of a slice of toast. She herself was rather petite, so it covered all the prime miking real-estate on her chest.

Here we go.

Luckily the necklace/manhole cover had a lot going on in its design. Stones, cut-out shapes, enameled sections, it was visually really busy.

Simple, hide in plain sight.

I just popped a black B6 through one of the fissures between stones, secured it to the back with tape, and let it just poke out into the breeze. There was so much going on in that piece of jewelry, that the tiny point of black just got lost in the dazzle. The weight of the jewelry acted like a mike stand and kept it solidly on her chest. She sounded terrific.

Sometimes even clothing is impossible to place a mike on.

One of the designers wore an extremely fashionable but revealing dress which featured just two very sheer strips of silk in a halter-top configuration which barely concealed her enchanting bosom. The sheerness of the material, plus the fact she was not wearing a bra, made concealment quite difficult. She did, however, have an asymmetrical haircut, shaved on one side, and thin dreadlocks on the other. These hung down past her shoulders.

I placed a black B6 secured with some black moleskin just above mouth level in her dreads. The amount of product in her hair, its dark color, and the fact it was braided all worked in my favor to make the hair a good foundation for the B6.

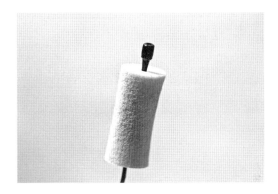

A Countryman B6 sandwiched
between two layers of black moleskin

B6 in a foam roll

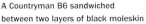

I was then able to run the cable from her hair to her neck, through the knot of her halter, and down her back to the transmitter attached to her waistband.

The B6 sounds best when it is in the open air. Small foam rolls with a hole in the center are an aid to mounting it between layers of clothing. Always make sure that the cap pokes out just a little. You can make these yourself or buy them pre-manufactured under the name of Hush Lavs.

As part of the sound team on a feature film shooting in a skyscraper under construction in Boston, we came up with a brilliant solution for miking the talent/construction workers. On a construction site, everyone must wear a hard hat. We took the actors' hard hats and installed the wireless transmitter in the top of the helmet above the webbing that supported the shell. The lavaliere, a side address TRAM, was taped facing downward toward the actor's mouth just under the brim on the front of the helmet. If an actor turned his head to the right or left, the mike turned right with it. Our mikes were always on axis. The transmitter was always up high for maximum transmitting clarity.

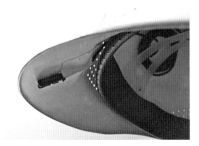

TRAM TR-50 under helmet visor shown without windscreen

There was no digging through clothing/overalls for mikes, and no clothing rustle. At the end of the day, we just collected all the helmets/transmitters, refreshed the batteries, and went home. We left the helmets wired up. It was brilliant. Simplicity always is.

A mighty wind!

You really can't protect lavalieres with the same zeppelins and softies that their bigger microphone cousins can benefit from.

We need to use tools that cover our tiny friends discreetly.

I was working on a documentary about David Macaulay, the author of *The Way Things Work* series of books.

The director wanted to do an interview on the beach with Macaulay sitting in a chair painted like a Holstein cow in a very wide shot. The wind was really blowing hard down by the ocean.

The TRAM I was using was getting pummeled by the constant wind, which was gusting to 30 knots. The little foam windscreen was overpowered. Crazy. In desperation, I took a cotton cowboy handkerchief from my back pocket, folded the TRAM inside and slid it under the author's shirt. It worked!

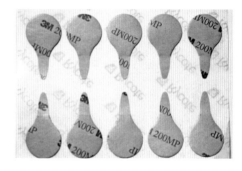

When your back is up against the wall, try anything!

One invention that has made life on location a little easier is a system by Rycote called Undercovers/ Overcovers, which consists of a number of parts, to protect one's lavaliere in the wind.

Starting with a sheet of teardrop-shaped double-sided sticky tabs, you stick the lavaliere on the surface.

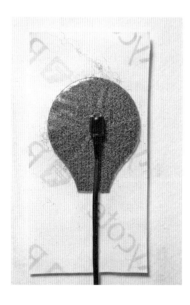

Then place a fuzzy, wind cheating cover (called an overcover) over it, which sticks to the same surface that is holding the mike.

Then take the paper off the back of the prepared, wind-protected mike and apply it to the talent.

The covers come in three shades for easier camouflage and two densities for various intensities of wind.

Once you make your tape/lavaliere/fuzzy stuff sandwich, you just peel off the back and place it on the talent.

Wind? What wind?

I recently covered an outdoor event where virtually everyone we interviewed was wearing a lanyard with credentials hanging from their necks.

If I placed the mike under their shirt, the credentials would rub back and forth across the shirt/mike and make a terrible noise. I took a Countryman B6 and the above combination and just applied it behind the wide strap of the lanyard. Then ran the cable, securing it along the way with Transpore up the strap to behind their neck, then down their backs to the transmitter on their belt or in their pocket.

Not just body mikes . . .

Wireless microphones can really come in handy if the director or cameraperson wants a super wide, and or, high shot. This could be a master for a feature or episodic TV show, or a documentary wide shot of a group meeting around a table. This is a typical scene shot to establish a scene or to give the editor options in cutting.

The TRAM has an attachment which solves this problem.

It is meant to turn the omnidirectional TRAM into a boundary layer/aka Pressure Zone Microphone. This configuration uses the flat surface of the table as part of the microphone itself. Once placed on the table or wall, diaphragm down, the microphone has a hemispherical polar pattern. It picks up sounds equally well in all directions in front of the surface, much like an omnidirectional mike.

By facing and micro spacing the TRAM's diaphragm against the table, the reflected sound delay is reduced and any interfering frequencies are so high as to be outside the audible range.

Another plus in this configuration is it increases the sensitivity of the TRAM. Since the direct and reflected waves add together in phase, the sound pressure doubles at the microphone, giving a 6 dB increase in acoustic level.

This is free gain. The microphone sensitivity increases 6 dB and the signal-to-noise ratio improves by 6 dB.

A TRAM pointed face-down into a table top

Conventional microphones' response to reverberant sound (room ambience) is weak in the high frequencies compared to direct sound. With a boundary layer microphone, the response is just as accurate as the response to the direct sound.

Boundary Layer-Hemisphere/omni placed on a table

Disguised with some books in plain sight. The transmitter is hidden as well

In a wide shot, the tiny black device will never be seen.

Hidden in plain sight

POINTS TO REMEMBER

- The use of multiple wireless microphones in reality television productions

- The importance of the boom mike in a world of wireless

- HOPS, the wireless link to camera

- How wireless transmitters/receivers operate

- The importance of systems maintenance

- How to troubleshoot and cure issues with wireless reception

- The world of the A2

- How to wire people in plain sight and under wraps

- Tricks with stash mikes

EXERCISE

Take five people and wire them hiding the mike in their clothing.

Listen to the results as they move and speak.

Repeat with a stopwatch. See how quickly you can do your job.

8 INTERVIEWS!

If sound were to have an equal to culinary's staple of bread and butter it would be the multi-angle interview. Interviews are in everything: news, documentaries, features, they are basically everywhere. Here we learn how to record the most fundamental of all scenarios, *but it is not as easy as it looks . . .*

The basic interview is done with one camera, a producer next to the lens asking questions, and someone being interviewed. In this situation, typically the interviewer's questions are going to be cut out/not used on camera.

The sound mixer (person), using a mixer (electronic equipment), with two tracks in a basic video set up will use a boom and lavaliere set up to mike one subject.

One microphone may sound better than the other, depending on conditions.

A lavaliere has an advantage in an extremely noisy room over a boom simply because it is closer to the sound source and benefits from the speakers internally generated bass via voice to overcome some of the annoying background sound.

In a normal/quiet sound field, the boom has the advantage with its more naturalistic response to the perceived distance of the camera from the viewer.

Do both. The talent may reach up and scratch his nose, hitting his lavaliere on the way and ruining the audibility of his response. The boom is unaffected. Back up.

Your obligation as a sound mixer is to not only listen to the quality of the sound but to how questions are being answered.

What do you mean?

For the sake of clarity and intelligibility, it is the producer's job to ensure each answer is a full statement.

If the interviewer's question is: "How long have you been working in this town?"

Subject's answer: "Four months."

Now, the interviewer's question, not on camera, not miked, is unheard and unseen. The editor has a shot of some guy saying, "Four months," without context.

Is that useful? Not so much.

The answer should be in a full statement:

Something like: "I have been working on this film four months."

It's not technically your job, but a producer may be inexperienced, or rushed, and may forget to brief the subject to answer in this slightly unnatural way.

A gentle reminder will make everyone look good.

Another scenario to look out for is *overlap*.

Let's say the subject answers rapidly and their voice is over the question, which will not be used, making both unusable. This is called *overlap*.

Pause, then answer . . .

Unacceptable noise intrusions

It is your duty in a real-life uncontrolled situation—say, while doing an interview on a city street—to, of course, be aware of intruding sound. However, it's only when it gets to be brutally intruding, like by a passing firetruck, that you should get the producer's attention and signal that the answer was just obliterated by a siren passing at 120 dB.

Sometimes producers get so wrapped up in the process they forget to hear.

It is really a judgment call. *Your* judgement call.

The worst thing you can do is just sit there and do nothing. Get the producer's attention, signal unusable (a serious shake of the head, a pursing of lips, a baring of the teeth, a finger drawn across the throat, an invisible noose held above your head as you mime death by hanging . . .).

Then, have the producer ask the question again.

Always get room tone

What is room tone?

Room tone is a recording of the atmospheric sound on the location where the interview, scene, shot, stand-up was filmed.

This is typically done on the conclusion of the interview/shooting. It is key to speak to the first assistant director or the producer/director informing them of this need to record 30 seconds of seemingly random quiet audio. If you do not immediately call for room tone at the conclusion of the shooting, people begin to leave, wrap equipment, make calls, generally assume that everything is over. It's not. Get the room tone.

Room tone audio tracks are key for the editor to complete their job in making the finished project seamless.

Room tone can be used to:

- cover odd noises during a pause

- bridge interview segments

- shorten answers

- make re-editing answers or dialog during cutaways easy.

On occasion I have called for room tone in the middle of the day. Make everybody stop, be still, be quiet, for 30 seconds. Just because. Just because *sound*.

The way two people are shot in *fictional feature films* is usually with the focus on just one person, reading lines for repeated takes with various inflections and degrees of intensity, from which in the editing room, the best take will be chosen for use in the film. Then, the cameras and lights turn around and get the reverse angle of the person the first actor was speaking to. They are reading a script, so all the questions and answers are covered. Everything is focused on one person at a time.

As a sound person, to cover this type of *coverage*, many sound mixers concentrate their efforts on the person on camera.

Of course they do! Why not? What other way could you do this scene?

Many mixers of feature or episodic television shows choose to mike the off camera talent who is feeding the lines to the actor on camera. This is recorded on an additional track.

Why?

Truth and authenticity of performance is one reason. Two actors, running a scene, no matter how many times they do it, are still genuinely responding to the other's tone, timing, and emotion. If the director/editor has a choice to use a complete interaction, no matter how brief, then they have a rare chance of an authentic and complete performance.

It is highly recommended as a mixer that you record dialog this way. Even if the non-visible actor's dialog is thrown away, you have still given the director a choice in post.

Isn't it twice the work?

Sure, a mixer may have to have a second boom, or activate the track of the second actor, but it really helps the authenticity and flow of a scene if you have as many complete interactions in dialog as possible.

There is a second reason for double-miking a single shot. Speed in post.

Many times in episodic TV, the editing department is under a time crunch to finish the episode. If the sound department provides both sides of an on-camera interaction, the editor can then spend time on other cutting issues.

The editor can also have the option of letting a shot run with the camera holding on the subject while their partner in dialog answers with no pause and full audio clarity.

Remember, the mixer's primary focus for audio is the person on camera. Any useful audio on the unseen actor is a bonus, so should any extraneous noise intrude on the unseen actor's dialog it is not considered worthy of asking for an additional take.

Record the off-screen actor on a separate track and they will love you in post.

The way things are done in a news magazine format show, like *60 Minutes*, *20/20*, or *VH-1*, is with two, or sometimes three, cameras.

- **Camera One** is an isolated shot of the correspondent who is asking the questions, and his audio is given a discreet track, say track one.

- **Camera Two** for the subject of the interview who is responding to the interviewer's questions and their audio is given a discreet track, say track two.

- **Camera Three**, if used, could be for a wide or tight shot, usually of the subject.

How do we do sound for two or three cameras?

First of all, we need to use a mixer with at least two microphone inputs: one for the interviewer and one for the subject. By its nature, a mixer implies that you are mixing more than one input, so we can safely assume that you have at least a two-channel mixer.

Second, decide how you are going to mike the subjects. Boom or lavaliere.

(If you are recording in multitrack, you can do both. More on this later.)

The booms are plugged into Input One and Input Two of your mixer.

They both sound great.

Why would you choose a boom over a lavaliere? Aren't booms more trouble?

It depends on what you consider trouble. The advantages of a boom are many:

If the room is acoustically sound, few audio mixers would deny that the boom sounds much better.

By better we mean it has a more "natural" sound, a more realistic spatial presence and can avoid many issues that come with using a lavaliere.

What issues?

A lavaliere, no matter how well designed, is out of your control, clipped to the body of another person.

They can inadvertently touch it, brush it, or move in such a way that makes contact with the mike causing a sound in your headphones akin to somebody tramping on your head.

If the subject is wearing awkward clothing, holding a baby, or gestures wildly, all of these can make a mixer's life miserable.

A boom flies high and out of reach of most issues of lavalieres.

With a boom, however, the subject may lean forward or backward unpredictably causing a diminishing in the quality of the audio signal due to the voice being slightly off mike.

This is an issue with mikes on stands.

If you have a boom operator, then they can follow the movements of the talent.

It has been my experience on most news/documentary shoots that most budgets will not allow for a boom person, hence the use of booms on stands.

A way to plan for such an eventuality is to choose a microphone with a slightly wider polar pattern for your boom mike.

A good cardioid, for instance, in a controlled situation can be as useful as a hypercardioid or supercardioid and allow for some subject/sound source movement without signal degradation.

An additional advantage, although more of a practicality/convenience issue than an acoustic one, is that the microphone is on a stand above the subject. When the subject enters, all they have to do is sit down. You do not have to wire them. You do not have to attach the microphone and transmitter, or the lavaliere power supply to their bodies and secure the cables against stress. When the interview is over, you do not have to jump up and gently remind them that you, like an audio pickpocket, have to slide your hands into their clothes to retrieve a microphone/transmitter/power supply.

Post-interview situations can be full of emotions, some not always helpful to the conclusion of your audio tasks.

I have had many an interview turn contentious; the subject became infuriated, and angrily stormed off of the set.

I have had to run after the injured/offended and politely yet firmly request my equipment back.

Awkward!

Back to our two camera shoot: correspondent and subject, both on camera.

Let's say you are suspending a boom microphone above each person.

The booms are on stands, regular light stands, fitted with a holder or cradle designed to hold a boom pole.

Now, to isolate one from the other, turn the pan pot above input one all the way to the left.

On Input Two, turn its pan pot all the way to the right.

The pan pot is the round knob with the L (left) and R (right) on either side of it.

Switch your headphone selector to left. You are hearing the left microphone. Tap on the right microphone.

You don't really hear anything.

Switch your headphone selector to right. You hear the right microphone really well, and the left just barely.

Switch your headphone selector to stereo. You now hear the left mike in your left ear, and the right mike in your right ear.

All good so far! Now let's plug our lines into the cameras.

You have an audio snake.

A what? A snake?

Yes, an *audio snake* is a cable that has multiple lines running through one insulated cable.

The most common one seen in field production has two balanced audio lines and one unbalanced headphone line.

So you can plug it into the headphone output of the camera and listen to the excellent audio you are feeding to your camera making sure no problems arise. A bonus to monitoring the camera's headphone feed will allow you to hear the two alarms that the camera will sound when either the media is low or the camera's battery is low and needs to be changed.

The back of a typical video camera

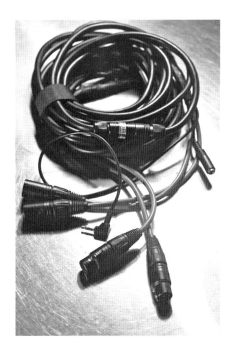

That's camera stuff. Why should it matter to me?

It allows you to plan ahead. You might want to adjust a boom or change a battery. You could stand by to hand the cameraperson a battery or some media.

Teamwork and *situational awareness* are really important.

I was shooting for a documentary that involved a famous modern dance troop. We were in their offices and my subject was at a computer composing a musical score. My sound man was a beginner. We were tethered together by audio snake, through which he was feeding me audio. I was about to move 180 degrees behind the composer at the terminal, I looked at my sound man and made a "C" shape with my hand gesturing my intentions. He nodded. As we moved behind the subject I heard a crash. My sound man had inadvertently knocked from the desk the Emmy Award that the troop had won. I briefly glanced down and noticed the globe of the Emmy intact, and thought, "Whew, glad we didn't break it." When I finally put down the camera, I saw the sound man and the producer struggling to pull the broken angel wings of the Emmy statuette out of the hardwood floor.

Something that represents the goddess of TV had just been broken.

OK. I only have one audio snake and there are two or three cameras. What do I do?

You need an audio splitter.

The type of splitter most commonly used turns one channel of audio input into three.

This is better than using a couple of "Y" cables because it uses a transformer and two separate switches to help suppress hum.

Isolated	Direct	Isolated
Output 1	Output	Output 2
Ground Lift	Input	Ground Lift

What do you mean, hum?

On a film set, there are a lot of cables with electricity running through them. They go to lights, monitors, and power supplies, and they are *everywhere*. When you run an audio cable across this spaghetti of power, you might encounter an *inductive current*. This is an invisible field of electromagnetic interference that can cause *hum*.

Hum is literally a humming/buzzing noise audible on the soundtrack.

The transformer in the splitter isolates two of the three lines of your audio signal, canceling out this hum.

The splitter also has two of its three lines wired in such a way to *lift* the ground. As we learned in Chapter 3 about balanced and unbalanced lines, a balanced line will work to eliminate hum; however, an intense hum could still intrude via the ground or shielding. These switches *lift* the ground by separating it from the circuit via the on/off switch.

Many mixers have two right/left-Channel 1 1/Channel 2 outputs of the same audio tracks.

Let's assume we have two audio snakes heading to cameras.

Make sure that your output levels are all the same. Line level or mike level, it's your choice, but they must be the same for both cameras.

Line or mike level? Does it matter what I'm sending?

Yes, it has to match what the camera inputs are set to.

If the camera is set to line, you have to send line.

In an interview situation, there are lots of lighting cables and with the risk of inductive current or electromagnetic interference threatening my tracks with hum, I will always send line level. The extra power of a line signal will go a long way to defeating inductive current.

I just plug them in, right?

A good sound person also has
to know about cameras. After all, the camera is your recording device.

Let's make sure that everything is ready for us to plug in.

There is a small audio panel on the side of the camera, or a menu page that designates audio routing.

Depending on the camera, you will be given a number of choices.

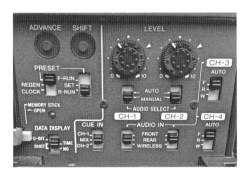

Usually they are:

Find the AUDIO IN area in the photograph.

1. **Front or camera mike:** This means the audio for the two main inputs is to come from the onboard camera mike.

2. **Wireless:** This means the audio is to come from the wireless slot on the top of the camera.

3. **Rear:** This means audio is to be accepted from the two rear XLR(F) inputs. Your inputs.

The last one is what we want. We are going to plug our signal into the rear/side of the camera. The selector switch for channel 1 and 2 should be set to REAR.

That is where you will plug in your snake.

Now go the rear of the camera and locate the two XLR (F) plugs. There may be a switch above each plug giving you a choice of either **line** or **mic**.

If your mixer is sending line level audio, switch both line/mic switches to line.

What does XLR (F and M) mean?

If the connector has prongs or a plug, it is considered male.

If it has a hole or slot, it is considered female.

The audio/electrical/scientific world is overwhelmingly male. Clearly, whoever came up with this system of determining connectors was a man.

Channels 3 and 4 can be fed audio as well. In the case of the camera pictured, they are set to receive inputs from the on board microphone mounted on the front of the camera, as a backup audio source.

Notice Channels 3 and 4 selector switches are set to F for Front.

This audio is useless for the interview: The on-board camera mikes are too far away, and too low quality. However, we will leave them activated, for should the cameraman be called upon to "Grab a quick B-Roll shot of. . . ." after the conclusion of the interview, and you are unable to get to him and the camera before he runs off (it happens) and/or he forgets to activate the Auto/Front switches for channel 1/2, the front mikes will still be hot on channels 3/4.

Insurance.

OK, one of your mikes is panned to the left, and one is panned to the right, remember?

Both mikes are reacting independently on the meters and you can hear each separately in the headphones when the monitor selection is set to stereo.

Back to the video camera.

We have found the two XLR (F) inputs and switched them to line.

Take the left XLR (M) of your snake and plug it into Channel 1, the XLR (F) on the left.

Now take the Right XLR (M) of your snake and plug it into Channel 2, the XLR (F) on the right.

Good job.

After plugging in the two XLR connectors into the back of the video camera, plug in the headphone return to the headphone out on the camera.

You might have to look around for this one. Camera manufacturers love to hide the headphone jack. The jack looks just like a headphone output on your mp3 player or smartphone.

Sometimes it's on the rear side, under a plastic flip-up door, or up on the handle. It depends on the manufacturer.

Turn up the volume for the headphone/monitor out on the video camera. I like it up all the way.

You want to be able to hear what the camera is receiving.

Again, you may have to search for the volume control.

It may be a knob right by the viewfinder, or a button on the top middle of the camera, or even in the menu.

On the camera pictured below, there are two knobs allowing you to control volume on the camera's audio either from a little speaker near the viewfinder (and the cameraman's ear) or from a 1/8th inch stereo phone jack where you have inserted the 1/8th inch stereo male phone plug from your audio snake. The knob above controls the volume of the alarm that beeps when the battery or media run low. Even when turned down all the way, it is still faintly audible.

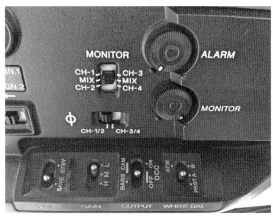

Make sure that the monitor select switch is set to what you most want to hear: in this case, CH-1 and CH-2 MIX.

This means a mixed two track signal of the two channels you have plugged your separate mixer inputs into.

On the camera pictured, there is an additional switch beneath the main monitor switch to give you the option of monitoring channels 1/2 or 3/4.

Now on the second camera repeat *exactly* what you did on the first camera.

Let's set audio levels on both cameras.

There is a switch, button or knob on your mixer that activates **tone**.

What is tone?

It is (usually) a 1 kH signal that is generated from the mixer to be sent via the outputs to the camera(s) or other recording devices.

Why do we use this tone?

We lay down 30 seconds of bars and tone on the beginning of the media to allow the editorial department to calibrate their audio levels and balance the color of their monitors.

This combination of test image and audio is to demonstrate the standard of the signal recorded in the field to which the technicians in the editorial suite can match the levels of their equipment.

Kill tone.

How can I be sure if I have hooked up the cables the same way for both cameras?

Have someone speak into a mike. Just one mike, the one you panned to the left.

Now have a look at the camera's meters.

Is the Channel 1 on Camera 1 showing audio? Good.

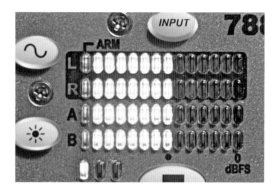

After sending tone, go to each camera and set your levels at −20 dB on both cameras

Activate BARS on a camera

Load the camera with media
Set timecode on the camera to 00:00:00 or 01:00:00
Roll the camera for 30 seconds
Repeat same on B Camera

Now, check Camera 2. Is Channel 1 showing movement? Good job.

Have someone else speak into the other microphone that is panned to the right.

Repeat your scan. Camera 1 should be showing action on Channel 2.

Camera 2 should have a meter bouncing on Channel 2.

Have some people sit in and speak.

Let both stand-ins have a normal conversation.

Set your headphones to monitor stereo.

Listen to the return on A Camera. Does audio in both ears sound good?

Listen to the return on B Camera. Good job!

If you are sending two channels of audio to the cameras and you are miking one person, then you have two tracks to play with.

You will want to split those two tracks or pan right and left.

Why should I do that?

By splitting the tracks, you are separating them. By separating them, you are isolating them from each other.

This is useful in post for a number of reasons, one of which is that should an unwanted noise—say, an inadvertent rub on the lavaliere by a subject—occur, it will spoil only one track.

If the camera is shooting just one person, and you do not need to mike the interviewer, you can put a lavaliere on the chest of the subject for channel one and use a boom above their head on the other channel.

Why do both?

The dual miking situation gives you and the editor choices. Many times a boom will sound better than a lavaliere. Should the subject brush their lavaliere accidentally, you have the boom as a backup. Should you have a dropout on the wireless lavaliere you are using, or the sudden occurrence of hum on a hard-wire lavaliere, there is the boom with its low impedance/high output hum resisting characteristic.

Always plan for trouble.

If you are doing a two camera-one camera on the talent, one camera on the correspondent, then you choose either two booms or two lavalieres.

What if I have only a boom and a lavaliere?

It will sound noticeably different, but depending on the voices of the players involved, you might be able to get away with it.

Say, for instance, both subjects have similar voices, like two 40-year-old men. The difference will be noticeable. However, if one is a woman with a higher-frequency voice and one is a man, the difference in microphone presence will be less noticeable.

What if I have three subjects?

Now you have to **mix** if you want things to sound great.

The correspondent is on one channel, and the two interview subjects are on another.

If two lavalieres are open, right next to each other on the same track, it will sound noticeably "airy." Not so good.

To avoid this, keep the non-speaking subject's level down to half of what the speaker's level is.

Most conversations have a natural rhythm. Listen. You will be able to tell when someone is going to speak. Ride the faders subtly.

You are now a mixer.

When I first started as a sound recordist, film shoots used Nagra reel-to-reel tape recorders. Awesome machines. Mono machines.

You had to mix up to three wireless and a boom during a take. You *had* to ride levels (MIX) to get your tracks to sound good.

What if there is a couch full of people and I have to get everyone?

First put the correspondent on one channel, say, a lavaliere/wireless. Then you have to boom, by hand, the entire couch. Remember if someone speaks unpredictably, the camera has to get there as well, find them, frame, focus . . . This is one time you have the speed advantage over the lens.

Talk to your cameraman, find out what his starting frame is. Figure out how close you can come in. Get a frame line (see Chapter 6).

Even if it is a two-camera shoot, the editor will usually go with the wide to record the beginning of the answer and then cut to the close-up camera for drama. It's an awful situation for everyone. I have had to boom bleachers, conference tables, the backs of trucks, a group standing on the deck of a ship underway in the Hudson River—you just have to depend on your natural sense of "Who is going to speak next?" and go for it.

Ugly TV, boomy sound . . . not the best situation.

Sometimes you have to respect the wishes of a subject regarding the use of a lavaliere or a boom.

I was about to record an interview with Dolly Parton, a revered star of country music, for a piece about her reading initiative foundation. As she entered the room, she saw the boom in place on a grip stand. She expressed pleasure at not having to be wired with a lavaliere, and asked if we could just use the boom.

Yes, ma'am.

The location was an insulated radio recording studio, so the boom would have sounded much better anyway. She was a charming woman.

INTERVIEWING IN A CAR

Cars can be a challenge. If the subject is the driver, with camera in the passenger seat, a lavaliere is not necessarily the answer. There are a couple of things you might be fighting.

The seatbelt/shoulder harness can ruin a lavaliere's day with rubbing, noise, and other mechanical noise against the mike. A car is a quiet place, especially a modern vehicle with the windows rolled up. Some situations demand a window rolled down, by directorial, or for the rear of the cameraman's camera to stick out of the car to get distance for a wider shot (wider makes the shot seem steadier against the motion of the car).

Then you have wind to deal with.

A well-stashed condenser mike is often the best option. If you have a detachable capsule system like the Schoeps Colette system, the system AKG developed with the 300B, or the Sanken CUB, you can place a mike just out of shot by the sun visor, buried in the dash, or in the center console. It just depends on the vehicle. Out of shot also makes a mike easier to windproof than a lavaliere in shot hidden in someone's clothes.

Even a boom right under the camera pointed toward the subject/driver can work better than a wind-whipped lavaliere.

The Schoeps Colette System allows the bulk of the microphone to be far from the receiving transducer/microphone capsule

Booming the back seat is easier. It often gives you more room and you are facing your subject rather than angling in obliquely between a pair of seats.

Just the thing for mounting in a car.

One mike is inside behind the Plexiglas divider on the passenger side.

The audio snake for this cable runs from inside the cab to the tow vehicle (Process Truck).

It is coiled in the middle of the hood waiting for the Process Truck to be connected.

Notice how the audio cable has been run in the crack between the hood and the fender. American cars are great for running cable both inside and out due to generous clearances in the way components fit together.

A taxi being prepared to be towed behind a Process Truck

POINTS TO REMEMBER

- The difference between a boom and a lavaliere

- How and why we should split audio tracks

- How to set up a camera for sound

- The importance of monitoring camera inputs

- How to record sound in a car

EXERCISES

Set up two microphones for a two camera interview using the knowledge from this chapter.

Once the cameras and microphones are set, and everything is working, send a classmate out of the room.

Disable two items in the record chain. It could be unplugging a microphone, removing a battery, turning down the headphone volume—really anything that creates an issue that holds up the production.

Get a stopwatch and start it when your colleague reenters the room. They have five minutes to troubleshoot the problem.

9 ON THE SET

"I think everybody should be nice to everybody."

Andy Warhol

There is a great deal of truth in Warhol's elegantly simple statement. Especially under the pressures and sometimes challenging conditions of a film set or remote documentary. There are, however, a few key additions to just being nice. They apply to pretty much any professional situation.

Consider the following points . . .

There are a couple of keys to success on set. They are simple, but every crew member I have admired professionally has practiced these steps. We all bring our best to work. If we didn't, we wouldn't be there.

What is a set?

That's like asking, "What is a restaurant?"

There is a huge variation on the type and nature of a film set.

It could be two people with a smartphone camera or a location covering acres on which hundreds of people are working.

Whatever the size or scale, some universal rules apply.

1. FOCUS ON THE TASK AT HAND
What do you mean?

You are not on holiday, not at a party, not at a bar, you are there to work. In your case, you have been hired to record some awesome tracks. Nothing else matters. Do your job. Keep focused even if the camera is not rolling.

Think about the following:

- What was the last thing you did?

- Was there a better way you could have done that task?

- What is coming next? Are you prepared?

- What could possibly go wrong/cause problems?

- Is the sound department ready for the action?

Think ahead: Is an external factor going to affect your workday (for example, are you going from interiors to exteriors, and does it look like rain?).

2. DO YOUR JOB WITH THE UTMOST PROFESSIONALISM
It is important to be social and friendly with your colleagues; however, there is a fine line between socially being part of a crew and talking too much. If production is not rolling at the time, talking about current events, sports scores, or telling stories is amusing and passes the time, but, in a room with many people concentrating on their tasks, do you really want to be a distraction? Nothing is worse than being the loudest guy/gal on the set, especially if the loudest person happens to be part of the sound department.

A loud chaotic set is a difficult and exhausting place to work. Sometimes on a reality set, I see mixers with a bottle of water stuffed into the front pouch of their rig. Drives me crazy. A stumble by the mixer into a solid object or the cameraman into the sound man's rig could burst the bottle, shorting out wireless transmitters or worse. Water + electronics = shorted-out/dead gear.

The real pro has a belt pocket/holder for their water bottle.

If you are working on a documentary, say, and someone with the best intentions offers you a drink, is it a good idea to accept it?

What if someone offers you a cocktail or beer? Are you sharper when you are sober or when you have consumed alcohol? Anything, like drink, that reduces your abilities you should stay away from. Have your drink after wrap.

Pilots have a rule: "Eight hours from bottle to throttle." This is sane advice.

Sometimes the subjects in your film demand you have a drink. This is problematic. You must be diplomatic.

I had just spent the day working under New York City with a group of rough-and-tumble men who dig the tunnels through which the subways and water flow to and from NYC. I was in a van cleaning layers of dirt and sand off my mixer with a toothbrush.

One of the principal "sandhogs" we had been filming appeared and aggressively demanded that the camera assistant and I come in and have a shot.

He had already had a few and was not accepting "no" for an answer.

I came into the bar, hung out for a few minutes, slapped a few backs, laughed at the day's events, and went back outside to work.

I was not ready for a drink.

Can you mix audio and eat a chicken wing at the same time? Only if you are dying of hunger should you work and eat at the same time. Energy bars are one thing, but food?

Like riding a bike and talking on your cellphone. You can do it, but is it wise? Are you devoting your full attention to both?

On the set of a documentary, you are often in someone else's space: their house, their business, their world. Respect that. The best way to exist in that situation is to blend into the woodwork. Be a quiet ghost. Do as little as possible to affect the existing mood. Try and change as little as possible of the mood in the room. Fit in quietly and do your job.

The campground rule applies when in someone's space. Do no damage, make no mess, clean up after yourself, pack out any dead batteries or other junk that follows film productions.

I once worked on the set of Clint Eastwood's *J. Edgar*. One of the quietest sets ever. A hundred people working, with no words being said unless they were absolutely necessary. The crew did their work quietly. No crashing stands, or tossed scrims. No open walkie-talkies.

Never did the assistant director yell commands. There were no bells, no warning, nothing but Mr. Eastwood saying "OK, let's roll." The crew was so in sync, that they knew when to roll, when to cut, when to pause, when to move on.

I had never seen anything so professional or a set that was as drama-free and tranquil.

3. SITUATIONAL AWARENESS IS KEY

What is this?

If you are a crew of two, do your job, but be aware of factors that will affect the outcome or progress of the shoot.

Like what?

Be aware of the level of remaining media in the camera. Do you need to change soon? If so, can you assist by having some media standing by?

Are the batteries in the camera beginning to reach the danger level? Can you assist in getting a fresh one, or assist in the changing out of the dead cell?

Is there dust or moisture in the air? Can you keep an eye on the lens and notify the cameraperson that there is enough flaky matter on the glass to bread a veal cutlet?

Situational awareness means you must devote your entire attention to what you are doing. Stop daydreaming. Don't text. Don't plan your next job. Don't organize the front pocket of your mixer. Don't snack. **Do your work.**

It's simple, really. Record awesome tracks.

Think about what else you could do to improve the sound. What else could you do to help the editor with this scene in the way of additional audio? If you could record this scene again, would you do it differently? How would you record it differently? This is how you grow as a sound mixer. By thinking things through, you become a more complete sound mixer.

When in the outside world, be aware of external factors that will possibly affect you and your team. Is that car turning toward us? Is the cameraman walking backward toward a busy corner full of civilians who might impede him or get into the shot? Is he walking backward toward a big steel light post?

Remember, the guy/gal that is paid to deliver the vision is actually half-blind with one eye stuck into a view finder. It's your job to look out for them, as you become the eyes and ears for two. Don't rely on producers to guide things; they are often too obsessed with content to realize even the most obvious hazards.

It's up to you.

Dangers can be subtle, like an uneven paving stone or approaching curb, or it could be a fire-fight that suddenly breaks out. Really, not even in a war zone. Deep in the historical section of Santo Domingo a crew of hundreds was shooting a scene for Michael Mann's feature film version of *Miami Vice*. I was headed downstairs from the shooting location to get the cameraman another camera battery. Two steps from the bottom of the stairs and the door to the street, a drunk security guard pulled a pistol on a Dominican National Guardsman armed with an M-16 assault rifle. As they say in the papers: ". . . there was a flurry of gunfire." I flattened myself against the wall. It came to nothing, no serious injuries. You just can't be too careful around guns, especially in a foreign country. See Chapter 1.

4. BE HELPFUL TO YOUR FELLOW CREW—WITHIN REASON

You are a team. If you are a hundred people or two, you can be considerate. If you are on a crowded set, and a grip brother might need someone to hold that net while they set the stand, offer. They will return the favor when you need a ladder or an apple box. If the grips are lifting a dolly on a truck, that might not be the place to jump in. They are trained in their jobs, you are not. Stay out of their way. Be polite. Hold doors for people. Hand things to colleagues that might just be out of reach.

It is a way of helping the production run smoothly, as well as demonstrating your professionalism.

5. TRANSCEND YOUR IMMEDIATE CONDITION

If you can't transcend your immediate condition, your work will become a record of your physical/mental state at the time you were working.

Let's say it's three in the morning and you have been working all day. It is very tempting to let things slide. It's easy when fatigued to not put in that plant mike, or extra wireless. When you are miserable from fatigue, lack of sleep, cold, hunger . . . that's when you really have to double down. Give it your all. Your colleagues can't see how cold it is, but they **can** hear wind on the lavalieres. They can't see how hard it's raining, but they **can** hear a wet muffled softie on a zeppelin.

I know some technicians that in the course of a production come to feel slighted or even cheated by the production. For this reason, they do less than a stellar job. Remember, *this is your craft*. To maintain integrity, to be true to yourself, you must always give 100 percent. It's their film, *but it's your work and craft, and ultimately it is you and your reputation.*

Rise above your cold/heat/fatigue/hunger/anger at the producers to do your best work.

While on set, you must respect the various departments. The bigger and more established the production, the more specific the departments become about slights, real or imagined.

It starts at the top, with the director.

It seems to me that the entire tone of a set is established by the director.

Some sets are tranquil, with abundant mutual respect and everyone taking satisfaction in creating a unique film experience.

Other sets are ruled by intimidation and fear with every department seemingly suffering under the autocratic rule of the director, who in turn passes the terror on through the first assistant director.

The best way to function is to do your job and to do it well.

Should a transgression between departments or individuals take place, real or imagined, immediately address what happened and work out the difficulty.

Life is too short to work under personal acrimony.

Ask the grips if you may have a six-step ladder or apple box before just taking one from their holding area.

Make sure the costume department knows before you are wiring a cast member, and allow them to be present and sign off on the job you have done placing the microphone.

If a producer visits the set unexpectedly, make sure you have a portable listening monitor available for them to hear your mixer's fine work.

In fact the distribution of personal listening devices (or IFBs) is one of the major responsibilities of the sound department on a feature or episodic TV show (see Chapter 13).

CLOTHING

Andy Warhol once said something to the effect that "Nothing is more embarrassing than showing up at a function wearing an inappropriate hat."

It's about the belt.

Really, it is. If you are new on the set and want to know who someone is or what they do, look at the belt they wear.

That woman over there, on her hips she's wearing an Ammeter, big pliers, heavy work gloves, an adjustable crescent wrench on a lanyard, and five or ten lengths of trick line dangling like scalps from her belt. She is an **electrician** (aka sparks).

That thin alert guy over there? On his belt he's got a big square pouch, a nylon cup holder in which is a can of dust off, a big leather holster holding a multitool, a flashlight, and a measuring tape. He is a **camera assistant**.

That wrestler-like looking guy? A big screwdriver, a set of allen keys, a socket driver, a box knife, and a ton of wooden clothespins clipped to his shirt. That will be a **grip** (see Kira Smith interview, Chapter 6).

The tall woman? She has a pair of cotton gloves, an IFB, a metal box with a bunch of wires running out of it, and a pair of headphones. **Boom operator**.

The guy with the big metal clipboard, a pen behind his ear, a long brown wallet looking thing on a chain in his back pocket, and not one, but two walkie-talkies and about four extra walkie batteries clipped to the very back of his belt. A worried look on his face makes all this add up to say: **assistant director**.

The woman in wool over there. She has a stopwatch and headphones around her neck, small flashlight with flexible neck in her pocket. A pen behind the ear and about ten more in her top pocket. She holds another big metal clipboard and usually wears glasses on her nose or around her neck on a lanyard. She also wears an IFB/Comtek. **Script supervisor**.

You can tell just by the belt and gear associated with it how together a crew person is. A set is no place for posers. It is a very mobile place where you often have to wear all your tools for immediate access. It might be a beautiful summer day outside, but inside on a dark stage, if you didn't bring your flashlight, it's going to be tough.

Common sense determines how we dress on set, but there are times and shoots that require forethought.

The photos of Hollywood in its golden age are filled with technicians wearing ties and vests.

Now it's cargo shorts and t-shirts. Crews all do pretty much dress alike on sets but on documentaries, you have to pay attention to the local customs.

You might find yourself in the midst of a religious service wearing a loud hula shirt. Or on Air Force One, getting a tough stare from the head flight attendant for wearing sneakers (which happened to me). Women should keep a scarf as head covering for that unexpected visit to a church or mosque.

I worked on a crew of five people in Irian Jaya, Indonesia. Three of the five came back with malaria. Why/Why not?

Use your head. Don't wear shorts and a tank top at sunset when the mosquitos come out for blood.

Check the weather. Access to rain gear is key. I always carry a disposable plastic parka in the bottom of my sound bag.

They are just big enough to cover me and my documentary rig from moderate rain/snow.

This rain thing astounded me when I first experienced it. I was working on *Law & Order* for a couple of days and our location was a pier in the Hudson River off the West Side of Manhattan.

I was walking out to location with my sound team and it was pouring. It was cold and we were miserable. We were getting just hammered by an icy March rain.

Jerry Orbach, one of the principal actors, came right up to me in the driving rain and welcomed me to the unit. I was nobody. Just a wet boom operator in a beret. He asked my name, shook my hand and won a place in my heart forever. What a nice man.

I learned from this that: *We are all in this together.*

I forgot about the rain completely.

When I looked at the episode, I was astonished: the rain was invisible. It just looked like a grey day in NYC. It's actually hard to make rain visible on film. It usually has to be backlit, or the puddle in which it forms has to be seen and lit to foreground the hammering of the drops.

COMMUNICATIONS

The standard form of two way communication on set is the walkie-talkie. Also known as walkies. Depending on the shoot many people wear walkies. It is very important to have all walkies equipped with earpiece monitors. More than one shot has been rendered unusable when from behind a flat, an open walkie blasts out a request from its speaker.

One challenging thing in the real world is that mixers must wear a walkie as well and not only listen to the fine work that they are mixing, but also monitor the camera channel to know what's going on.

CELLPHONES

I shouldn't really have to say this, but all phones should be considered and turned off. Alarm/timer mode on the phone should be off as well. The worst thing is when the sound person's phone goes off. It does happen.

Once, when working an intimate interview, my cameraman's phone began to bark. Yes, bark. He thought the silence switch would stop all sound output. All but the alarm.

All visitors should be warned about silencing their phones. Often it is a well-meaning outsider who commits this most embarrassing gaffe.

BELLS

A brief explanation of bells:

On some sets/sound stages, the sound mixer or one of the production staff will activate a series of lights and bells (ringing buzzers) to keep the unwary from entering the set during a sound take. This is known as "lock up."

To call for the bells and lights, one of the production staff calls; "OK, lock it up!" Just like a ship leaving dock to enter a channel, there are three short blasts on the bells, and the red flashing light or sign is switched on.

When the take has concluded, and the set is again open, there is one short blast on the bells and the warning lights are switched off.

Two blasts are used to lock up for a rehearsal—meaning continue to work, but very, very quietly.

POINTS TO REMEMBER

On set, you must:

- focus

- do your job

- be aware

- be knowledgeable of the big picture

- be helpful

- transcend your present condition.

The relevance of the bells:

- three blasts for shooting lock up

- two blasts for rehearsal lock up

- one blast for set released.

---

10 PREP IS EVERYTHING!
Your Basic Kit

". . . whatever can happen, will happen."

Augustus De Morgan 1866

"Everything that can go wrong, will go wrong."

Captain Edward Murphy USAF 1948

What follows will help prepare you for the inevitable. Things break, go flat, lose power, get forgotten, fall through the cracks, which ever expression seems most adequate for what will happen: failure.

It is part of life, and especially part of a technical reality. You must be ready for it. We will spend the next chapter counting the ways of dealing with failure and fixing what broke.

You are working on a dream job. As part of a three-person team, you are filming a documentary in the South Pacific about Amelia Earhart. Imagine flying aboard a Cessna Caravan over the vast ocean somewhere south of Howland Island, one of the last locations Earhart's plane was thought to have been bound. It's not so dreamy today, however. A big storm is pushing your single-engined aircraft around like a leaf, possibly off course. The electrical storms crackling around you have made communication between your pilot on his HF radio and civilization impossible. As you look over his shoulder to watch him enter frequencies on the radio the entire control panel goes blank and the engine warning lights flash red.

Lightning strike!

In the course of working, we all face lightning strikes of some degree.

This chapter is to prepare you for the worst, and to make sure you have the proper backup to deal with most situations.

Clearly, the odds of having to use your audio kit as an aid for lifesaving is remote, but when your tools fail, or are effected by external forces, the shoot must go on!

When something breaks, you must have the presence of mind to troubleshoot the situation logically.

The pilot of your Cessna Caravan in a descent over the Pacific is trained to go to his **engine out checklist**.

The aircraft manufacturer and his fellow pilots have prepared a list of actions to take when power is lost. It's the same on a film as in an aircraft's cockpit: *time is of the essence!* Just as the pilot watches the ocean get closer and closer, if you have a problem and production is waiting, then the odds of you being re-hired are getting more and more remote . . . as a professional you must be able to *access the situation* and then test and try A, then B, then C . . .

In the following chapter we will cover:

- what you should carry as minimum equipment in your basic kit

- common malfunctions and how to troubleshoot them

- what to do when a piece of gear becomes inoperative and how to circumvent disaster and continue shooting.

What is in your basic bag/kit (minimum gear to shoot) is largely determined by what kind of film you are working on. While all films are different, let's use the video documentary kit as an example. Although the gear may vary, the philosophy is pretty universal.

The film business is an unpredictable one. It does not matter how large or small the production is, things change at the last minute. Weather, schedules, and events all make life very unpredictable. The only thing you can control is your gear. From a shoulder bag to a van load of equipment, *you can control that*. Control and order amidst chaos can exist in your sound package. Get everything just right and keep it that way.

Plan for failure

First of all, before we talk about individual components, let's look at the overall threat of not being able to shoot at all.

I have learned, the hard way, that if you are flying anywhere to a job, **always carry your essential core kit on board the aircraft with you. This means everything.**

Everything for one day's shooting.

This includes your core gear as well as batteries, media, boom, cables, and so on.

Luggage gets lost or diverted. It usually takes a day domestically, or many days internationally, for your tools to find their way back to you. Even if you are flying from New York to London, say, and your kit gets mishandled to Dubai, sure, production can obtain you a rental package to use that first day, *but that takes time*. If you're bound for Algiers? Not so simple. Carry it on, it won't get lost. It's a drag hauling a big kit of stuff through terminals, security, etc., but thousands of bags go missing every day.

It's the professional way to travel.

What about that bag?

My bag is a large, nylon shoulder bag made for still photographers, with many padded compartments. Choosing a bag is like selecting a wallet. It's very personal, chosen to suit your own needs. One bit of advice: whichever bag you choose, *get the next size up.*

You will be surprised at how quickly you will fill up space.

Designate pockets for specific items and stick to that system

If it's dark, if the bag is between your legs in a van, or if it is stored in the row behind you in a theater and you can only reach but not see it, you have to know where to go.

The things you are going to need most often are your expendables. Like media cards, main power batteries, small batteries for transmitters, etc. Determine what are they are and where they should live. Make them easy to get to.

For instance, I keep my CF (and SD) cards in a small zipper pocket in the very top of the main bag flap. I know where they are, they are easy to get to, and I can access them in total darkness.

Additionally, if you have to send an assistant or P.A. for something vital, you know exactly where, and can tell them which pocket to dig through, it will expedite your equipment needs.

What else is in the top of my bag?

I reserve another pocket for semi-audio items of great importance.

These include spots for:

- a flashlight

- credentials/press pass

- a small container of sunblock

- a small container of bug repellant

  - *Not just for me, but for my boom mic windscreen. For some reason, bugs love to circle my microphone when I'm working. A normal-sized insect circling inches away from a condenser microphone sounds like a dentist's drill.*

- a couple of pairs of earbud earphones for cameramen to monitor audio or producer/director to use for their IFB when they have left their headsets in the van

- a tiny multitool device (pliers/two types of screwdrivers)

- a US to EUR AC power plug adaptor

- a tiny compass (never had to use it; I just like knowing it's there)

- green tea bags. The importance of caffeine cannot be overlooked.

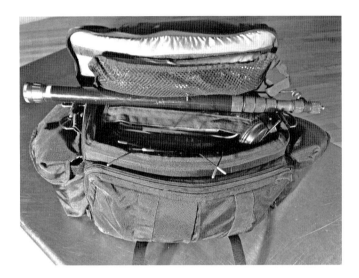

Top of bag, second level

Beneath the rain flap that has all the above stuff, this is a long zippered compartment.

This is where I keep the main XLR cables:

- one 15ft XLR cable

- one 25ft XLR cable

- the main two channel snake with return cable link between mixer and camera

- one 15' BNC-BNC for timecode jamming.

Net zipper pouch beneath the above folding compartment

This is where all my adapter cables live:

- headphone splitters

- TC Lemo to BNC for jamming timecode

- XLR shorties

- other general helpful but not immediately urgent small cables.

A spare coiled boom cable, the cable abused the most, is there as well.

The main compartment

This is where the **mixer/recorder** fully built with wireless receivers rests. This is the heart of the audio package. It's protected on all sides from moisture, sand, dust, sun, and anything else that may harm it.

I am of the philosophy that you leave the mixer fully built. Pull it out of the bag, slap a battery on it, plug in the boom, and you are ready to send audio to camera.

Camera just needs a battery, media, white balance, focus (or not), and it can shoot.

The sound person has a much more complex set-up.

Simplify your start sequence. Some situations won't allow for an elaborate assembly process.

Be ready.

Big front zipper compartment

This is one of the easiest-to-reach pockets in the bag.

It contains:

- my arsenal of microphones, mic mounts, and wireless transmitters

- the vital AA/9v cells to replenish the transmitters with power

- commonly needed tools, like a flashlight, small screwdrivers, selection of moleskin/molefoam, surgical tape, windscreens, etc.

Fast access is everything.

Right side pocket

- The big windscreens for the boom mic, and all my barrel/phono adaptors for converting one thing to another.

- Occasionally a clamp to hold a boom atop a light stand lives here as well.

Left side pocket

- A plastic fishing tackle box which contains:

  ○ a store of lavaliere mics

  ○ mounting aids (pre-cut mole foam/skin hushlavs, windscreen material, mounts: vampire clips, tie clips, button hole mounts, PZM mounts).

- A selection of elastic belts to hold transmitters securely to the talent's body. In black/white/flesh: belt straps, thigh straps, and ankle straps.

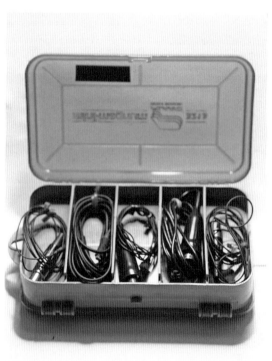

- XLR to TAF 3 cables in mic and line for obtaining audio feeds from other sources.

Rear zipper pocket

This is the most accessible of all. Accordingly, this is where the spare main power batteries live. Also a great place for reading material.

A brief word about fasteners:

Velcro = noisy = bad.

Zippers (especially nylon) **= quiet = good.**

An example:

You are in the pit right behind Loren Maazel while he rehearses a full symphony orchestra. You are laying down an amazing stereo mix, plus tracking the wireless he's wearing for his instruction to the players.

What do you think will happen when you open that long Velcro flap during the pianissimo portion of the program?

Game over. Thanks for playing.

COMMON MALFUNCTIONS AND HOW TO TROUBLESHOOT THEM

Power

As we have seen in Chapter 4, the "cockpit" of your audio world is the mixer. You find a spot, hook up to the camera, set your boom, prep your transmitter/lavaliere, and go to the power switch of the mixer . . . nothing.

- Are the batteries fresh?

- Is the power switch selected to the correct input? (INT/EXT)

- If you are using an external source, is the cable connected/or broken?

- If you are using internal cells only, is the battery door closed or making proper contact with the cells?

If all else fails, try shaking the mixer from side to side. Often in well-used mixers, the battery contact springs lose length and will not make proper contact with the cells.

Power on/no audio

The power light is on, but I hear nothing over the headphones.

- Try activating tone. See it on the meters? Is it in your headphones?

- If yes, then it's an input/mic or monitoring problem.

- If no tone in the headphones, try the A/B switch. This allows you to monitor the mixer or directly from the camera. Switching it to the mixer should allow you to hear the tone.

Still no?

Be sure the headphone level is up.

If the camera is off or its monitor is down, or the headphone input on your audio snake is out, you will hear nothing.

Tone but no input audio

Everything is on, but I can't hear the boom!

Center the pan pots, switch the headphone monitor to stereo. Still no joy?

Trace everything from the microphone back to the mixer.

• Plug the mic right into the mixer.

Working? Yes. You have a bad cable.

Working? No. On the mixer:

• Check the input at the XLR. Is it line or mic?

• Line: you will hear nothing. Switch to mic.

• Mic: still nothing. Check the mic powering switch.

   Set to your mic's needs: 48v or 12v.

   Tap it. *Does it register on the meters?*

   It does? You still can't hear anything?

   Check headphones for short.

Having a bad boom mic is rather uncommon, but possible.

One of the most common cables to break is the coiled boom cable. It suffers the most abuse, always getting pulled, yanked, and caught in car doors and other things as you move around shooting.

Always carry a spare.

Always monitor audio from the camera/recorder.

Things could be just fine in your headphones as you monitor from the mixer. However, if your outputs don't match the camera's inputs, you are blindly sending distorted tracks.

If you are sending line output to a camera that is set to mic, you will see levels during tone that look good. Unless you listen to the output of the camera, you will never be aware that you are sending distorted garbage that is totally unusable.

Additionally, listening to camera will allow you to hear any hum, or other issues that might happen in either the camera or your multi-cable snake that you are feeding audio through.

The information panel on a video camera showing two channels
of incoming tone at −20 dB

Hum. I hear a hum.

What is hum?

You know, the buzzy noise you hear when plugging in speakers to your computer.

Now imagine that constant hum all through your audio.

Can you say unusable?

WHAT CAUSES HUM?

We learned in Chapter 9 about ground or grounding a circuit. If you have a constant hum, it is
possible you have a broken cable whose ground has been interrupted causing electronic
dissonance.

- Check your cables by connecting and re-connecting them, or replacing them one by one. Plugging
  and unplugging, testing the connections, is another strategy. Perhaps a connector is not seated
  properly.

- Wiggle the cables at the connectors. Breakage usually happens at the cables' connectors. Find it.
  Replace it.

- Look to see if another piece of electronic or lighting equipment is nearby. Power cables can produce
  an inductive field around themselves. Keep them away from your audio lines.

  ○ If it is necessary to cross power lines and audio lines, do so at a right angle. This will expose less
    of your audio line to the nasty power currents.

- Often while using hard-wire lavalieres the line leading to the power supply is balanced but the
  microphone cable itself is not. A recipe for hum. (We learned about the difference of balanced and
  unbalanced lines in Chapter 2.)

  ○ I personally never use hard-wired lavalieres. The reasons?

1. to avoid hum.

2. to avoid hum (yes, it's that important).

3. interview subjects often forget they are wearing a microphone, stand up, and begin to walk away, damaging the thin microphone wire or connector. I will put a wireless mic and lavaliere on sit-down interview subjects every time.

REALLY? What about risking a hit on your wireless?

Wireless interference: *I'm getting hits!*

A "Hit" is when the audio from the wireless transmitter is interrupted, or "drops out."

Hits can also take the form of static, burst of noise and other interference.

If you are rolling, what steps can you take?

• Move. Get closer! Move physically closer to the transmitter with your receiver.

• Check the transmitter's battery. Low power can compromise a transmitter's performance.

• Check the condition of the antenna on the transmitter. If it is not seated correctly, or broken, it can compromise your transmission.

• Look around for any physical complication to receiving your transmission.

• With walls, chain link fences, machines producing electrical fields (motors), get away and establish a line-of-sight path between you and your transmitter.

• Move the transmitter on the talent. If it is buried in their waistband, and they are a person of substance, their own body can suck up or block your transmitter's radio waves. Move it higher, if possible, and with as little between you and your transmitter.

• The ultimate cure: find another frequency that is not compromised with interference. If you have the option, do a frequency scan and find some nice, quiet bandwidth to move to.

Don't you get interference or hits during an interview while using wireless instead of hard-wire lavs?

Pretty rare. I mean, I'm only a few feet from the talent. I did my freq. scan. Fresh batteries.

It almost never happens.

Hum with a hard-wire lavaliere?

Much more common.

MAYDAY! MAYDAY! THINGS AREN'T WORKING!

When you have a problem, and you are the sound person, you are pretty much *on your own.*

No one on set can help you. Usually no one is aware of your issues. You are alone under the headphones.

When a lighting instrument blows a bulb, everyone can see that. Those present are usually patient as the electrician replaces the head or the bulb.

If you have an issue, no one can see it. Patience under the pressure of a tight schedule is hard to come by.

You are like that pilot heading toward the blue Pacific.

Don't panic. Troubleshoot the problem. Use your experience. Analyze the situation. Try one thing at a time. Move efficiently.

Especially when you are just starting as a mixer, you want to appear confident, professional. That can go out the window when a snag comes up you cannot solve. When you are stumped, have an expert you can call.

We all rely on our colleagues, especially when you have tried A, tried B, and still, it's not working.

Take a minute. Pick up the phone, and rely on another mind to shed some light or lend a fresh skill set to the problem solving process. We have all been there.

<u>**Inform production immediately if you have an issue that will adversely affect the sound**</u>.

If you have a problem during a take, *by all means: speak up*. No one knows that it was unusable except you.

Never ignore an issue that will cause problems later.

It's OK. It's not your fault. The camera team is always getting another shot.

Demand another take. It is your right.

If the director says "Don't worry about it," that's their call.

You did your job informing production of the situation, but by all means *let them know*.

LIGHTNING STRIKE!

Let's return to the scenario at the beginning of this chapter. The plane begins an oddly quiet descent. Through the windshield you can see a tiny island no larger than a football field.

What do you have in your kit that you could possibly use to hasten your rescue?

What is important here is your troubleshooting ability in addition to your kit that will aid your rescue.

The plane you are flying in has an Emergency Location Transmitter (ELT) just behind the passenger compartment.

This, upon activation by either impact forces or manual activation will transmit a distress signal to a satellite overhead every 50 seconds. The ELT runs out of battery power after 48 hours. It is up to you to make sure it keeps working, and it is transmitting properly.

Say your pilot ditches in the surf just inside the barrier reef around the small island. After safely relocating ashore, remove the ELT from the aircraft to rescue it from seawater, and check that it is transmitting. If the ELT antenna atop the fuselage is broken, remove it and with the aid of your 25' XLR cable and some adapters place it securely atop one of the palm trees.

Short of a Satphone, that's your best hope.

At the airport, while waiting for the flight, you picked up a local island newspaper. Go to the TV listings for the local UHF station. What is its channel? TV 31. What wireless systems do you have with you?

Block 21 and Block 22.

Block 22 is 563.700 MHz to 587.500MHz. Look at your frequency card. That's TV25 to TV32. Tune a transmitter to 575.500 and another to 577.900. Both are TV 31.

Record a short message on your SD card recorder stating emergency, who you are, and where you are (coordinates from the cameraman's GPS). Set the SD card player to repeat and put the whole bundle—transmitters, SD card recorder, and the correct patch cables—in a waterproof bag and up against a tree.

A long shot, but better than a note in a bottle.

You have tons of batteries, but let's say this is your home for an extended period of time. The ELT battery is good for a while but, what else can you do for power?

Get the wooden pallet that the freight from the broken Cessna was loaded on. Pull out the galvanized nails that hold it together. Pick up a ton of those bitter little limes that are all over the ground. Place a nail in each. Run wire between the nails. Wire them in series. You have just made a battery. Test it with your multimeter. When you get 1.3vDC, that's enough to power a transmitter.

Just as you were imagining cutting squares of aluminum to hang from the tallest trees/to reflect sunlight/to signal passing ships, the Cessna's control panel awakens. The pilot has reset the panel breaker and the plane now begins a slow and steady climb . . .

He kept a cool head, took the right steps, and solved the problem.

So will you when the time comes . . .

EXERCISE, PART I

Pack an equipment bag with everything you need for a one-camera shoot. Leave the room.

Have a colleague remove *one key item*. Now, set up for your imaginary one-camera shoot.

- What are you going to do to make the shoot happen?

- What can you substitute/rig to make up for the missing item?

Now it's your turn. Repeat the exercise with your colleague as missing item-victim.

Repeat, and repeat until you have at least three discovery/solution problem solving drills completed.

EXERCISE, PART II

Now that you have your equipment set up perfectly and operating flawlessly . . .

Leave the room. Your colleague is going to alter a setting, change a switch from mike to line, switch off mike powering—in general, screw something up.

It is up to you to test for all systems and find and correct the problem, all in five minutes. Switch roles and repeat.

---

11 SYNCHRONIZATION

In the twenty-first century, one thing we take for granted is the linking, or *Synchronicity* of sound and picture. It was not always that way. In fact it is one of the great technical achievements of post World War II film technology. Even today, unless you are very careful, issues could arise that will cause major headaches. Here we will learn about *frame rates* and *timecode*, and hopefully stay out of trouble.

What is that board they hold up in front of the camera that has all the numbers?

That is a slate. It has been around, in one form or another, since the beginning of film.

Why does it have a stick with stripes on the top that swings down and makes a noise?

When film first began, a slate was important to identify each shot. It shows the production company, the date, the scene, camera roll, and other information important to the production.

The stick on top with the stripes was added once sound began to be recorded on film sets. It is used to visually synchronize the soundtrack with the picture.

This is done by swinging the board down, which impacts the top of the slate, making a loud noise.

The noise is recorded on the soundtrack.

Later, in post-production, the editor will look for the film frame that has the top part with stripes matching the bottom set of stripes as they come together. The editor then finds the sound of the impact of these two pieces of wood as they come together and matches it with the above image. The editing machine is locked and the sound and picture run as one.

We say the sound is *synced*.

In the event of a scene starting unexpectedly, or for some reason without a slate, the slate is photographed/recorded at the end of the take. This is called a "tail slate."

The procedure is, once the on-screen action is over, to hold the slate upside-down in view of the camera, say "tail mark" or "tail slate" clap the sticks together, then rotate the slate from upside-down to upright and hold for a beat so the information on the slate can be photographed.

The upside-down slate indicates to the editor that this is a tail mark not a head slate.

What about those numbers running along the front of the slate?

The numbers are timecode, or SMPTE timecode.

What are they? Why do we need them?

Timecode looks complicated, but it is really just a very precise counter.

What does it count?

Timecode counts every frame of video the camera shoots. It assigns each of these frames a number.

These numbers are based on one of two things: **the passage of time, as in a very precise clock**, or a **running count of the frames as they are shot by the camera**.

The first type, based on a running clock, is called **free run timecode**. It's just like a big precise stopwatch that never stops. You can set it to start it at any number. Most productions start the clock running to match the time of day. In a documentary this is helpful in post to find shots, especially if you know when during the day of shooting they happened.

The second type of timecode starts when the camera begins to record and counts each frame until the camera cuts.

This is called **record run timecode**.

This is video. Why do we, as sound people, need to know about it?

The audio person, typically the most tech-savvy person on the set, has always been the person responsible for sync.

SMPTE timecode was invented in 1967, for television, but would not become relevant in production sound for another two decades.

What is SMPTE?

The **Society of Motion Picture and Television Engineers** is a US-based, international professional association of engineers working in the motion imaging industries. An internationally recognized standards organization, it has for almost a hundred years established standards, recommended practices, and engineering guidelines for television production, filmmaking, digital cinema, and audio recording.

Back on the set we begin to record audio and the counter (SMPTE timecode) starts. The audio recording medium has SMPTE timecode imprinted on it. The camera photographs the numbers on the slate for reference, to mark the start of a shot or, in some cases, the end of a shot during a "tail slate."

The editor matches the number on the audio medium with the photographic image of the number on the slate and locks the edit machine to sync the footage (matching precisely image and sound).

The timecode also gives the script supervisor precise start and end points of scenes to aid in the editorial process.

At the beginning of the day, the sound department powers up its recording device equipped with a timecode generator. It begins to run with the correct frame rate for the production.

The sound team "jams" a slate's internal timecode generator, or clock, so it matches the timecode from the internal generator on the audio recorder.

What is "jam"?

This just means taking an output cable from the timecode generator and plugging it into the receiving device. By pressing a switch, the receiving device will "jam" or match up exactly with the host timecode equipped recorder.

Now you have two very precise clocks running exactly the same time.

The slate is remotely mirroring the audio recorder's time signature on the recording medium.

The same time signature that is being recorded on the audio media can be passed precisely to devices other than a slate.

You can take a cable and feed this time signature to a video camera. Advanced film and video cameras have precise motor controls that drive the medium at constant rate.

That camera, if set correctly, will automatically lock frame to frame with the timecode from your recording device. Now your recorder, the slate, and the camera, are all running timecode in sync.

In review, the audio department master timecode signal "sets the time" on the camera, the slate, and often an external sync generator (also known as a sync box or Lockit Box) to be attached to the camera by "jam syncing" each device.

This is as easy as plugging in a timecode output from the audio recording device, and powering up or pressing a jam switch on the receiving device.

So from our recorder, using timecode, we can take countless devices: recorders, cameras, slates, and have them locked together *to the frame*.

Standard procedures mandate that all devices, with fresh batteries, are synced at the *beginning of the day before the first shot and again after lunch.*

SMPTE/EBU/AES all have standardized a number of timecode frame rates.

What is SMPTE/EBU/AES?

These are all regulatory organizations that set standards throughout the industry.

We already are familiar with SMPTE.

EBU: The European Broadcast Union is a union of broadcast stations/organizations throughout Europe. Sort of a United Nations of television. One of its missions is to promote standards between its various members to ensure seamless interaction from broadcaster to broadcaster. Pretty much all European countries' national broadcasters belong to the EBU.

They are also responsible for the **Eurovision Song Competition**, which we really won't discuss much in this volume.

AES: The Audio Engineering Society has a membership made up of scientists, engineers, and working professionals involved in the professional audio industry. The AES is the only worldwide professional society devoted exclusively to audio technology. It is involved in the formation of standards, much like SMPTE, for analog and digital audio recording. Its convention features a dizzying collection of microphones and audio systems that border on the (audio) pornographic.

This is the line-up of timecode in our time:

23.976/24p

Currently the most popular is 29.976. It is also referred to as 24p (the "p" standing for "progressive"). This is a frame rate used by film or video productions that are planning on transferring a video signal to film. Film and video productions use 29.976/24p even if they are not transferring to film, simply because the on-screen "look" of the low frame rate resembles traditional film.

In the 1930s, the speed of 24 frames per second became the standard speed for motion pictures.

Before that, film speeds varied from 16 to 18 frames per second. This explains why when footage shot from this era is seen today, projected at 24 FPS, things seem to move in a comically fast and jerky fashion.

The images should look jerky, for the film is being projected at a faster rate.

24 FPS

This exists for time only applications where no video/digital transfer will take place.

Not surprisingly, this mode is used very little; it is an artifact of the past.

25p

This is another progressive format that runs at 25 FPS. This is due to the European PAL standard of 50 interlaced fields per second. There are two worlds: countries that use 110vAC/60 Hz as the electrical wall current, and countries that use 220vAC/50 Hz. This rate is derived from the PAL television standard of 50i (or 50 interlaced fields per second in broadcast television).

Productions for the parts of the world that use 220vAC/50 Hz electrical current use this rate for compatibility with the 50i television fields.

29.97 (non-drop)

This rate was developed due to the fact that color television frames do not actually run 30 FPS but slightly slower. This slowness was originally to accommodate a color signal within a channel originally designed for black and white. In this format, every 30 frames the timecode "counts" as one second. It is actually slightly slower. Because of this slowness, it takes more than an actual hour of time. So, in short works this is insignificant, but if the program is an hour long, it actually runs 1:03:06. This is a big deal to a network running longer programs.

29.97 drop frame

In answer to the +3.6 seconds of the above format, 29.97 was invented so that the time could closely match clock time. No actual frames are dropped, just frame numbers are skipped to keep on clock time.

30 (non-drop)

This is used for shooting commercials or music videos for TV, ignoring the 0.1 percent timing error as irrelevant due to the shortness of the piece. Not used much these days.

30 (drop frame)

This is used in shooting television long format or episodic shows. This format, once the standard, has largely been surpassed by 23.976/24p. The delivery master of the show is intended to be at 29.97.

When the film and tape are transferred they are slowed down by 0.1 percent, becoming 29.97.

Continuity

Timecode is a really practical and efficient way to maintain sync throughout a project. The key word here is "throughout." For ease of workflow and continuity from production to post-production/editing, and on to delivery, the producer/film maker must have a clear understanding of what timecode is used during each phase.

As a freelancer/hired gun sound mixer, it is your responsibility to obtain timecode frame rates before the shoot from production.

Conversely, it is the responsibility of the production to specify this key information.

If you have not received these specifications, you must reach out to production or if they are technically challenged, get them to call the editor to get the specs before the shoot begins.

Often the sound recordist/mixer is the most technically advanced member of the team.

Your craft, by nature, demands a more advanced baseline of knowledge. You are aware of issues that other departments are only vaguely aware exist.

In demonstrating your awareness of the larger picture, you can bring great peace of mind to production team members who may be facing these issues for the first time.

If you are making your own film, then make life easy for yourself. Choose a timecode and stick with it all the way through. You will thank yourself in the edit room.

One element of slating that is invisible is the recording of sound effects. These are sounds recorded to be used later in editing the soundtrack of the project. They are also known as "wild" sound. The wild refers to the fact that they are not linked to picture by sync, hence wild, as in running wild.

You can ask for and record dialog from an unseen actor that is off-screen, from a telephone, or is meant to repair a brutal sound take later in post. These are known as wild lines, and are slated verbally by the number of what scene they should accompany.

If it is a sound effect, unrelated to dialog, it will be slated by using a take number beginning with 1000.

On would say for example: "Effect 1001, room tone for scene 140, the garage, 30 seconds starting now . . ." and roll sound for 30+ seconds.

Setting timecode on a video camera

At the beginning of every shoot, the mixer/recordist addresses the issue of timecode.

Many shoots set timecode by time of day.

What is that?

Well, time of day is just that: the time at the moment, when the production is taking place.

You look at your watch/phone and enter the time of day.

Sony F10 Video Menu Screen

How do I do that?

Look at the lower part of the panel that has all the switches. Look in the upper left-hand corner.

Right below the rubber button marked SHIFT is a switch. It allows you to PRESET timecode to anything you want.

Move it to the center position marked SET.

The numbers indicating hours will begin to flash just like when you set your bedside alarm clock.

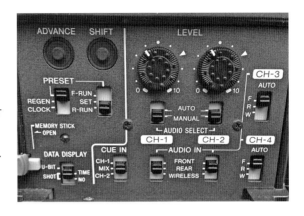

Set the timecode in a 24-hour clock mode 1 p.m. would read as 1300, 2 p.m. would be 1400, and so on.

You set the numbers by pushing the two rubber buttons marked ADVANCE and SHIFT.

When you have found the correct hour value, push SHIFT and the blinking number will move to minutes, and so on.

Once you have set the time, move the switch up from SET to F-RUN. F-RUN stands for FREERUN.

The clock will begin to automatically count up frames, seconds, minutes, and hours. This camera has now become the master clock.

You can jam other devices, usually cameras, from this camera's TC OUT socket on the back or rear of the camera. The type of connector used is almost always a BNC type connector, which is the standard video connector.

A BNC connector

You take the other end of the connected BNC and connect it to the device that you want to jam by attaching it to the TC IN BNC socket on the device to be fed timecode.

On the host device, say another camera, go to the identical panel you have just left on the host camera and again in the upper right-hand section, just beneath the button marked SHIFT, move the switch to SET, then to F-RUN. The timecode you have just set on the host camera will appear on the display above the control panel.

Disconnect the jam BNC cable and watch the numbers. Are they still holding? Great. Always check with the cable disconnected after jamming.

Record Run timecode

This is used when, typically during a multi-camera shoot, the powers in post ask for Record Run timecode.

Then the sound mixer sets the timecode on the "A" Camera to 1:00:00:00.

Each camera other than "A" Camera is set on Free Run instead of Record Run.

BNC cable is run from "A" Camera's Timecode Out to the "B" camera's Timecode In input.

Once they are connected, the "B" Camera begins to mirror exactly the timecode running on the "A" Camera.

Then on the "B" camera a BNC cable is run from Timecode Out to the Timecode In input on "C" Camera.

"A" Camera becomes the *master*. "B" and "C" Cameras become the *slaves*.

"B" and "C" Cameras receive mirroring timecode from "A" Camera through a daisy chain, or interconnecting series of BNC cables.

All cameras are locked up, frame to frame, with timecode.

The procedure is: once the cables have been run, the mixer, having run their audio lines to each camera, sends tone to all cameras.

Media is in place in each camera.

With tone coming in to each camera, and each camera's levels set at −20 dB, bars are switched internally to display on each camera.

This means color bars are appearing in each viewfinder/monitor.

Then, on command, each cameraman rolls at the same time, and watching the timecode counter reads aloud the seconds on the timecode display to assure everything is synced up . . . "51 . . . 52 . . . 53 . . .," chant the three shooters in unison (or whatever numbers the code reads) until it is clear that sync is obtained.

They roll for 30 seconds. Thirty seconds of bars and tone.

Why bars and tone?

Bars and tone were originally used to allow a reference for the editor in post to set their studio video monitor levels to match the field material color reference and audio levels. In other words, the editor would play the 30 seconds of color bars and tone. They would set the edit monitors to match the bars on the shot footage. They would set their audio levels to match the −20 dB tone sent and recorded on each camera's media.

Then they were ready to edit, making confident, informed decisions about the material they were cutting.

How did picture and sound sync up before video?

That's a great story . . .

In the early days a second camera would run along with the picture camera and record sound on 35mm film. The audio information would be converted into light impulses. These impulses would expose the film with waves of light that corresponded to the sound waves being picked up by the on-set microphone.

Since both were plugged into the wall/house current of 110–120 volts AC, at the same time, they had frame-to-frame sync.

This might have been OK for a studio, but the whole thing was really large and cumbersome. Shooting sync sound on location was awkward, to say the least.

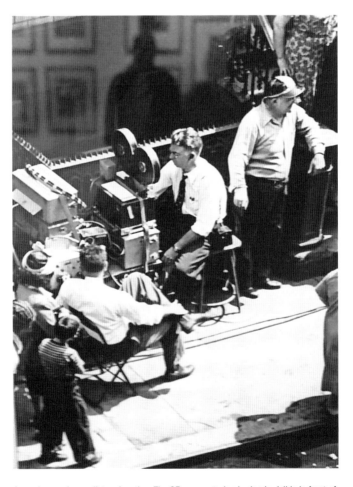

An early sound recordist on location. The 35mm magazine is clearly visible in front of the mixer.

A clapper slate was used to sync up the sound with the picture after both were processed. This was done on an upright Moviola editing machine.

With the widespread use of magnetic recording tape after World War II, recordings were made in the field using large reel-to-reel recorders using 16mm full coat magnetic stock. The size and power consumption needs of these recorders made them less than easy or practical to use. In the 1950s, Eastman Kodak released single perforation 16mm motion picture film with a magnetic stripe down one side so you could record directly to the film as it ran. The slow speeds made sound quality for anything but speech an issue. With analog recording, the faster the magnetic tape moves, the better it sounds. Tape speed is directly related to the amount of noise that a recording will have. The faster the better. A film camera at 24 FPS is slow indeed.

Some cameras also had optical heads which allowed an optical soundtrack to be imprinted on any single perforation film stock that was not magnetically striped. This was a sort of miniature version of what is illustrated above with the 35mm optical film recorder. Again, quality was telephone grade for voice only, due to the limitations of the medium itself.

An example of a rather large RCA 1950s vintage location magnetic sound-on-film recorder

Kodak introduced 16mm film that had two magnetic strips, one for recording and another thinner one for balance.

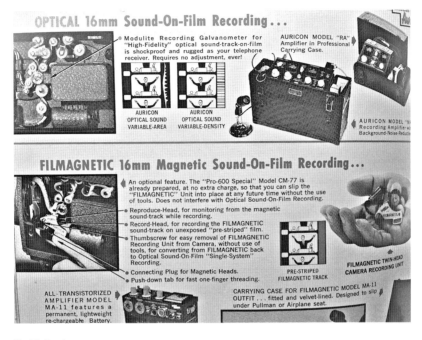

The Big Break

Without a doubt, the greatest advance for sound in film was an invention in 1951 by a young Polish engineering student in Switzerland.

Stefan Kudelski

The Nagra II (note the crank to wind the spring tape drive)

Stefan Kudelski patented the Nagra I, a shoebox-sized reel-to-reel tape recorder that equaled the quality of most studio recorders. Swiss radio stations were his first customers for his improved Nagra II.

Kudelski's Nagra III, developed in 1958, was able to be synced with film cameras, by the use of his neopilot recording head

The neopilot head, which utilized wave cancelation, made a recorder, weighing in at around 14 pounds, a versatile and high-quality instrument. This freed the sound mixer from the heavy and complicated systems of the past.

With such a tool, documentarians were able to make films with just a two-person crew, allowing an intimacy unknown in past film productions. Locations could be anywhere, freeing film makers to improvise, or in the case of Direct Cinema, create documentaries following subjects virtually anywhere.

By the early 1960s, Nagras were standards of the industry. From the 1960s to the 1990s, virtually every film was recorded with a Nagra. Not only were they incredible recording devices, they were built to be virtually indestructible.

Production Sound Mixing

The Nagra recorded a 50 Hz or 60 Hz signal, the same crystal frequency that controlled the motor of the camera, in to the recorded sound. It was never audible, for the neopilot head was really two heads in one recording the sync signal *out of phase*. It was *"heard"* by another dual neopilot head *only in playback* when it was being resolved in the transfer to 16 or 35mm magnetic film stock. A brilliant system, really.

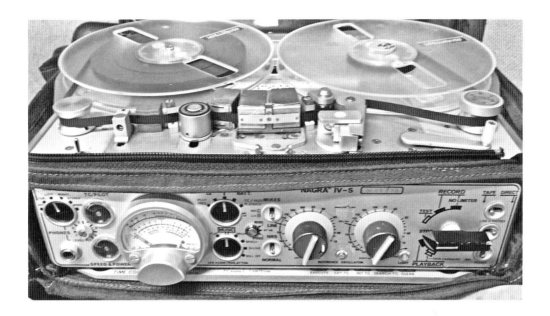

Nagra introduced timecode by adapting their Nagra IV-S machine, recording the code between the two stereo tracks.

A most un-user friendly machine, but beloved by those who mastered its operation system.

With the introduction of timecode the medium became less important as audio technology began to work its way through various video based linear formats like DAT (Digital Audio Tape), Tascam's DA-88, and various DVD-RAM machines before settling, at the time of writing, on hard drive and solid state recording medium. The machines and manufacturers that embody this technology are:

• Sound Devices 700 series, the 702/744/788 and their mixer/recorder combinations the 633/664

• Zaxcom's Diva (discontinued)

• the Nomad.

In the days of the Nagra, one 5" roll of 1.0 mil tape would last 22 minutes. That's two 10-minute 400' film magazines running at 24 frames per second.

A compact flash card, depending on the size, will provide the recordist with hours of recording time.

You can carry in a normal pocket a week's worth of recording media if need be. Amazing.

POINTS TO REMEMBER

A production must have sync at all times.

Always slate.

If you think sync has been lost, immediately note this and work to re-jam and correct the lapse.

Tell the script supervisor or the producer. Call or have them call the editor.

EXERCISE

Take a timecode slate, and jam it from your recording device. Disconnect and observe its precision.

Now set the slate to 30 FPS and the master timecode on your recorder to 29.97 FPS.

Observe how quickly the numbers begin to differ.

12 POST-PRODUCTION: Where Does It Go From Here?

You have recorded a lot of tracks you are proud of, and some that were disastrously bad.

You hand in your media. Then what happens? How do you explain what went where or how you did this for that? We will talk to two sound professionals, a mixer of episodic TV shows and a master of cutting dialog for projects with modest budgets. What they have to say is gold.

In this chapter, we will see how sound recordists interact and communicate with the editors: we will talk to John McCormick about his work in episodic TV and dialog editor Alex Noyes about his needs.

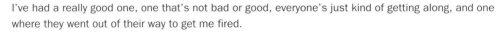

INTERVIEW WITH JOHN McCORMICK

John is a New York-based sound mixer. He is a veteran mixer of long running episodic TV shows such as *Unforgettable*, *Rescue Me*, *Royal Pains*, *Third Watch* and *Law & Order*.

How have your relationships been with post?

I've had a really good one, one that's not bad or good, everyone's just kind of getting along, and one where they went out of their way to get me fired.

The good one was Ilene Horta, the post supervisor on the Dennis Leary show *Rescue Me*.

We had a very good relationship because she was approachable, you could call her, leave her a voice message, give her a heads-up if trouble was on the way. You could call Ilene and say:

"Look out, you are going to get a piece of crap."

She would get back to you and say:

"Oh yeah, it was bad, but we did this, and that . . ."

You just didn't want to surprise her if you could avoid that.

Give me an example of that.

It would be a very noisy location, an actor not talking very loud . . . the thing is on *Rescue Me* . . .

Dennis Leary would not go to the ADR. So if Dennis Leary wouldn't go to ADR then neither would any of the other big actors. Only on rare, strange occasions. Now Leary would give me a wild track if I needed it, but we're recording sound in mock fires, but they are real fires, so you have a fairly noisy background. Once in a while you would lose some dialog in a place, or you want to get a piece of tone to help something out but couldn't, or you get hammered by a plane and they won't do another take . . . you would just make that call:

"Ilene, this is going to be pretty dicey."

We once had a script that called for two fire trucks to answer the same event, Dennis Leary's fire truck and another rival fire truck competing for the same call.

Production Sound Mixing

There were 14 or 15 speaking parts in that scene and a production mixer has at the most eight channels. So I went to the director and asked about how he was going to cover this.

He said:

"I really can't tell you."

So not knowing, I was able to go to my unit production manager and my producer and say we have these 14 people talking and I can't handle them all, so what we are going to have to do is hire an extra production mixer.

This is before we were using the Sound Devices 788. We were recording to disk on the Fostex 824 Eight Track Recorder.

The new mixer, since he didn't know any of the players anyway, I gave him the other fire truck, and I took Leary's since I knew the actors. So we had we had one boom person and all these radio mikes.

He recorded eight isolated tracks of each of the wireless mikes his fire truck crew were wearing. From his mixer he did a mono mix. He did not record this mix but sent me a feed of that mono mix so that it went on one of my eight tracks. I did a mono mix as well, and recorded it on another of my eight tracks. The boom was isolated on a separate track and the rest were isolated wireless tracks.

When Ilene got the original DVD recordings of the day's shoot, there would be two mono mixes and all the rest isolated tracks of each of the radio mikes and the one boom.

I called Ilene before we shot that and said:

"Guess who's coming to dinner?"

She said: "What do you mean?

"Well, you are getting two mix tracks of 14 wireless and a boom all on two DVDs with matching timecode. This is sort of a Big Magilla."

She agreed.

The interesting thing was that she wasn't even on the clock yet.

This shoot was early in our schedule. Sometimes your post-production, the very people you deal with, if your start you show in July, they sometimes don't start work until you get closer to the mixes.

She came in on her own time, listened to it, and called me back and told me, "It's all there, It is interesting, the mixes are terrible, but we got some good stuff, there are some problems here and there, whatever."

Ilene was really good, she knew her job, she was aware of our challenges, because her dad was an old sound effects guy.

I went out to LA for a mix at the end of our season once, and I watched her plug her Powerbook and drive into the incredible digital complex and pass off all of these soundtracks with timecode and everything to the post mixers: the dialog guy, the sound effects guy, and the music guy. She gave them all of that stuff and then went to another stage to an ADR session for some other show, and then she came back just to see how it was going. It was stunning to see.

Ilene was the high water mark of post.

Post people need consistency.

That's not always possible. You might get a guest director for an individual episode, and they have different agendas than you and I as sound mixers. You will have days, or an episode where you have someone that is not very kind to sound. They can really put a divot into what the post people are expecting to get on a daily basis. That's when I try and talk to the director and say: "We normally hit this mark better. I need this. They are expecting this level of quality from me."

You try and win the battle that way.

Give me an example of how a director would not be kind to sound.

The director would keep shooting with two cameras, a wide shot and a tight shot at the same time.

He wouldn't turn off the wide shot to let you get the boom in for the close-up sound.

The acting isn't so much in the wide shot, it's in the close-up; that's where most of the action is going to happen.

It's understandable where wide and tight sometimes must be shot simultaneously, but in one case, there's an actress wearing lingerie so there's no way to wire her, and we are inside. My boom person can't be close enough. I have a plant mike taped to a closet door angled at her the best we can.

I'm getting a track, but it is not the quality I normally hand in.

The director just had to have his way all the time.

It wasn't such an issue on the street with multiple cameras shooting wide and tight shots. I couldn't get a boom in but it didn't matter because I had wired each of the firemen's helmets with the mike just under the front brim.

I joked with somebody from props:

"You could shoot them from the Space Shuttle and I wouldn't care."

They replied:

"Don't suggest that. He will want that shot. . ."

Why do you do a mono mix? Don't you have all those tracks isolated?

Then what would the editor cut with? And what do the people with Comteks/IFBs listen to? The post people are good, really good, but they're not there on location. So they use my mix to at least get a sense of what's going on.

Or if you have two actors with two radio mikes and they are standing in a glassed-in doorway someplace, that's a little reverb-y or whatever, there are any number of times that the person who is speaking, when they are speaking, I'm using the other person's lavaliere. When they quit talking, I'm using the other person's mike, because they sound better. It just sounds better. Would a post mixer think to use the other person's radio mike? I don't know that they have the experience of sitting in the hell-holes that we sit in.

Do you do sound reports?

I do keep paper sound reports. But I do work for other shows filling in for other mixers, that have gone away from paper to generating a report on the Sound Devices 788. I take the CF card out, put it into my laptop, and rename the sound report and then e-mail it to the editor.

What kind of notes do you put in the sound reports?

Ninety percent of the time it is noting something informational, a plane, a wireless hit, some lavaliere rustle.

If there's something that's helpful, don't just leave it in the sound report, because they don't read the sound report all that much, but if you tell the script supervisor something is helpful, they do read that.

For instance just a week or two back, we had rolled sound, slated, and then we were waiting for an ambulance to pass before we could shoot. As it turned out it happened to be very quiet, there weren't a lot of crew people talking. So I put on the report: "Very good city ambulance tone, in case somebody wanted to yank it off and use it." So I ran over and told Valery the Script supervisor: ". . . that was a really good one." Then we had the same thing with a helicopter a little later in the day. A sound effects person would not even know that it's there, but it is, and it's easy to get to.

A director can be really helpful holding a shot after the dialog is finished if there is a minimally intrusive noise like a plane or siren that is not so loud as to keep us from recording. A jet plane or an ambulance sound that is so far back that the dialog is above it if the director lets the shot run, you now have the perfect piece of tone to lay over the other side when you turn around for coverage to smooth out the mix. That tone is perfect, because it's right there and it's from the shot, and if they don't say cut and everybody keeps quiet, then you have it.

How do you give your sound to production? Where does it go?

Now we are just handing off these CF cards [Compact Flash].

It goes to two places that you have to think about.

It goes to the place where it gets synced up.

After that it goes to the editors, and then to the post people: dialog/effects/music.

It's good to have the phone number of the place that does the audio layback (syncing) if for some reason camera/sound goes out of sync you could let people know.

As you get closer to the mix, then it's good to talk to the editors. I do talk to editors. If not an editor, then the post-production supervisor.

Find out who they are, put them in your cellphone and check with them once and a while.

"How's everything going?" That's all.

Also, when the show is on the air, it's nice to watch the show, if you can, and then call back and complement them on their work, because they do nice work. They take our tracks, blend them together, add effects, and then the music, they save dialog pieces and make it work. We are just a piece in the jigsaw puzzle of a TV show.

Why would you recommend being a sound mixer?

You and I did all those years of documentary work, and I think that was the greatest extended education that a human being could possibly have, hanging out with smart people, being with all different kinds of people from all over the world and getting to listen. Having a microphone in a room and hearing stories. Just think of the different kinds of things we have worked on . . . There's that.

Feature work, now that's fun because you get to work with actors, and you are part of this huge team of people.

Jerry Bruck once said about sound people in the movie business, and I recall his quote:

"Sound people on big feature films are really just the feather on the ass of a very large bear. And it shall be sat upon at various times."

Wait, this is fascinating, but I just have this little short film that I want to sound really good . . .

All right, you have done everything with great care.

- Your successful location scout resulted in intelligent decisions about places to shoot.

- You have chosen the right tools for the job, and prepped them to function perfectly.

- You have studied the script and to the best of your ability, made decisions about recording tracks that really can make an interesting and clearly defined film.

- You have documented your work in notes and metadata to make identification and synchronization a simple and clear proposition.

Now it's time to pass off your fine work to another skilled craftsman.

Let's talk to one of the premiere dialog editors in NYC.

INTERVIEW WITH ALEX NOYES

Originally from Washington DC, Alex began his career influenced by experimental music, tape manipulation, early synthesizers, and Antioch College workshops at MIT where he found himself the youngest, a student, amongst the head of music programs of major universities. After working with Norman Seaman, concert agent, Alex worked for Harvest Works bringing his skills for the first time to film/video. Later at Mercer Street Sound Alex worked with numerous independent filmmakers mixing fringe art pieces and documentary films. From this Alex branched out into early video game soundtracks.

What's the most helpful thing a film maker can do when he walks through the door of your mixing studio?

Typically, a film maker/director who is also the editor, will walk through the door with their edit files: OMF or an AAF depending on the editing system being used, and a set of sound notes of what their ideas are.

It's really useful to sit down with a filmmaker and watch it through their cut and create essentially spotting notes.

This is where you get a real idea of what it is they are really trying to go for.

You learn what problems need to be solved and what the filmmaker's intent is.

What problems need to be solved and where you can suggest things that might be different approaches than what the direction they have been headed in.

The importance of having other people's eyes on a project is essential.

The great sound editor Walter Murch once explained that he does not like to be on the set of films he was to mix because he felt that he wanted to be viewing the film from the audience's point of view.

The sound editor can be a contributor to the feedback process during shooting or not. It depends on the director.

A lot of this is interfacing with the picture editor.

A lot of picture editors don't think about organizing their tracks according to sound.

A very important thing is constancy in terms of track organization in terms of production tracks.

If you have booms on specific tracks and lavs on others, stick with that.

Stick with your channel assignments throughout a production, which can be hard if a film has multiple sound mixers.

Documenting stuff and passing that along makes that process that much easier.

Getting sound logs from people these days is rare but incredibly useful.

I've worked on a lot of films where there are no sound logs at all.

"Here's my hard drive!"

What I see a lot of these days is they do folders for the day of production.

That's not really enough.

Many films currently are edited with a nonlinear program like Final Cut or Premiere and the director will walk through the door with what they feel is a pretty complete film.

How do you handle that?

Sometimes people come in and they don't hear any problems when they're editing.

And they think everything is OK.

That's because they are editing on laptop speakers or with small speakers in a noisy edit room full of air-conditioning and hard drives and they can't hear the sound.

Then they come in to work with me and they listen through my full range system in a reasonably quiet room and they are like: "I never noticed that sound before, we've got to fix that."

How/where do you start?

The first thing I do is start splitting out dialog tracks so I can get an idea of what the contents are of each of those tracks.

If I am the sound editor and mixer, which is very often what I do, basically all of the post-production sound, then the process becomes sort of fluid, you aren't just mixing and editing you are doing a little bit of everything as you go along, the first thing I'll work on is dialog.

I'll ignore everything else because the number one thing people should be concerned about is getting a good dialog edit.

That is basically what makes the film work.

As much as music is a wonderful tool in filmmaking—and with sound effects you can have all sorts of things going on and create interesting soundscapes—making the dialog seamless is what makes the project coherent.

Number one, that means it's really important to capture that stuff and capture it in a way that will make it work in the final production.

Especially with lower budgets, ADR is not an option.

Editing dialog involves using a lot of room tone creating fills and cleaning up things, finding and loading in alternates for takes, and cheating where you need to cheat.

That's basically what a dialog editor is all about.

What I like to say is that 50 percent of the time you are trying to create continuity where continuity does not exist. Then that other 50 percent you are doing the opposite: creating perspective where perspective does not exist in the track.

What's the biggest sin someone can commit?

First, you have to stress to your sound mixer/boom operator/cameraman-sound-one-man the deal is that <u>if they have a problem, they have to be able to stop and deal with it.</u>

Make sure they know they have permission to be able to do that. So that you don't wind up in an impossible situation later on.

The other crazy thing to say is that you have to listen to the recordings that you are making as you are recording them, you know?

I think that should be obvious, but I've been in situations where people set something up and assume everything is fine and feel that they don't need to monitor the recording, with tragic results.

The other thing I would add to that is that location scouting and thinking about sound when you are thinking about locations is the other thing that I think is under-appreciated.

Before you start shooting you have to decide that this is not a good place to shoot our scene because of sound.

As much as we like what it looks like it's not going to work.

I would love to see scripts.

Look at the script and see potential areas to empower sound as a storytelling element.

As much as you talk about visually storytelling, which is a powerful tool, you can do the same exact thing with sound.

There are thousands of examples where sound can lead the story and explain what is going on. and that's what people don't think about when they are beginning a project.

You can look at that and find it in the script. That is the place to start.

That's in a narrative project. In a documentary you can do almost the same thing by looking at the big picture and letting sound enter that big picture.

How should a young filmmaker who knows nothing about sound approach audio for their first project?

There are all sorts of examples that they can go out and find, of really using sound, it's all throughout film history.

I guess that's number one. It's there, you just have to go and find it.

I think there are filmmakers who aren't generally focused on sound. If that's the case then they should find someone who is and bring them into the project right away.

Whoever. It could be a musician who is a friend. **To open their ears. Start to think about active listening.**

One of the things I love about what's going on now is an interest in soundscapes and acoustic ecology. I hope that starts to influence filmmakers some, because it think there's a lot of interesting ideas there.

About production sound, and this is a really big change in the way things are done, and it has a lot to do with the newer portable sound recording systems, the quality of the microphone systems out there now, you can really capture your sound on location, capturing the ambiences in the spaces you are recording in, capturing individual sounds, for production sound effects, and if you spend that additional time in that original space it is easier and saves so much time in post without having to resort to recreating sounds through Foley.

POINTS TO REMEMBER

- The workflow of a major film production must be considered.

- The importance of communication has been stressed over and over.

- A good relationship with your colleagues in post is paramount.

- Clear documentation of audio tracks is always useful.

- Consistency, both in the field and in post, is a virtue.

- A good dialog edit is what drives the film.

- Allowing the sound mixer to stop and fix a problem is key.

- Listening carefully during recording, and in life, cannot be stressed enough.

EXERCISE

Take a short film script. Rewrite it in a way that the story is told only through sound.

Just sounds.

When finished, go through it and add another level of sound to add subtlety to the first layer of storytelling sounds.

"Empty the pond to get the fish."

—ROBERT BRESSON

13 OPPORTUNITIES FOR SOUND WORK IN FILM/VIDEO

"We often miss opportunity because it's dressed in overalls and looks like work."

Thomas A. Edison 1883

What do you want to do? Within the world of sound one can have wildly different experiences. From a documentary sound person, running through middle eastern streets surrounded by gunfire to mixing choirs and organ in a cathedral. You are really using the same information/intellegence to achieve the same goal: Create great audio tracks. We will survey the possibilities of how one can have a life through sound.

In this chapter, we will look at the different paths you can take as a sound person for film/video.

The world of professional sound recording is vast and the opportunities are many. For this book we will concentrate on just the world of sound for film and video.

If you are interested in studio music recording, a great book that presents a historical overview as well many great audio truths is George Martin's *All You Need Is Ears*. Martin was the studio producer for The Beatles from their beginnings at EMI and during their most creative period, when they produced Revolver, Sgt. Pepper, and Yellow Submarine. It is a great record of audio history in the making, as well as being full of recording techniques helpful to all sound mixers.

If you start with a tiny pile of gear you have scraped together, then you have a great knowledge base that you can build on, and obtain vital troubleshooting skills. I mean, if you have only one mike cable and it craps out, you have to fix it, right? Miss dinner, sit up in your room, and learn how to solder on a tiny bedside table, to be ready for the next day. Priceless.

ONE-MAN BAND

I always kind of hated this expression, but accurate it is. You must shoot, record sound, and perhaps do a little lighting, strapping everything to a little hand truck, and dragging your load all around town by yourself.

What you will need

Depending on your mission, you need a basic kit to record sound. Let's say you have an actual video camera with XLR inputs, tripod, a camera mounted light . . .

What kind of sound gear will I need?

The camera usually has a microphone mounted on it. It may be a good professional-quality shotgun or a mike that came with the camera of dubious quality.

Either way, when shooting, unless the subject is right next to the lens, and you are in a dead quiet environment, a camera mounted microphone is of minimal use at best.

Take the next step, and place a lavaliere on the subject. You may have a hard-wire lav or a wireless mike on the subject. Plug it into one input of your camera, set that input to mike level and either manually set levels as the subject speaks or place the camera on automatic level control.

Essentials

• Hard-wire mike set up

• Two lavalieres, one to use, one for backup

- If battery powered, fresh batteries and backup cells

- Two 20' XLR cables: one to use, one for backup

- Headphones to monitor sound

- Wireless mike set

- Wireless transmitter/receiver

- Plug-in lavaliere, backup plug-in lavaliere

- Fresh batteries, plus a set of extra fresh cells

- A short XLR-XLR to attach the wireless receiver to your camera (and a backup cable)

- Headphones to monitor sound.

Handheld microphone

Often in this kind of situation, a handheld mike will be necessary if you have a correspondent to ask questions of subjects. Remember most handheld mikes, dynamic or powered, will need to be held close to the subject's mouth to get that sweet broadcast quality sound.

The dynamic mike will need an XLR-XLR to go from the microphone to the camera. It will be plugged into an input set to mike.

Wireless handheld microphone

This set up works the same way as a wireless lavaliere, except that instead of a body pack transmitter and a plug-in lavaliere microphone, the handheld microphone will have a plug-in transmitter that "plugs in" to the end of the microphone.

The receiver will plug into the camera input set to mike. The plug-in transmitter, the one that fits into the end of the dynamic mike is often referred to by a slang term. This term is an unfortunate but commonly used expression with which you should be familiar. The plug on a transmitter is often referred to as a "butt plug." Beware of using this term in sensitive company, like at an interview with the Queen.

Back up

Did you notice the pattern in the above list of needs? Back up. Back up. Spare. Back up.

Cables go bad. Especially if the gear is from a rental house, it may have developed a short, and not been marked as such. Staff gear? The same thing. Staff crews often do not care for the house equipment as if it was their own. An extra cable or cables, although a weight and storage liability, will be so worth it when the one you are using shorts out or begins to hum like a beehive.

With batteries, you can never have too many extra cells.

I have had fresh cells installed in a piece of equipment the night before, fail the next morning due to one new cell failing, leaking, and crashing out due to a manufacturing defect. It will happen.

Be ready. Nothing like backup gear to make you confident.

TWO-MAN CREW/DOCUMENTARY/REALITY/ENG

Now you are a team. As you move and shoot handheld, you and your cameraman are dance partners. As in dance you can have a spooky partner who is unpredictable and steps on your toes, or you can have a dream that you automatically know before when and how they are going to move.

What equipment is used

The cameraman usually has a large "serious" video camera with two XLR inputs for you to input your awesome audio mix/tracks.

You, the sound person, have a mixer from two to eight inputs, and mixes/distributes the audio you are mixing to the camera and possibly recording internally if your mixer is so equipped.

You must also feed two tracks to your camera.

How do you get your fine work to the camera? The most common and simplest way is via an *audio snake* (see Chapter 8).

This is a thick cable that is made up of two balanced XLR lines and one unbalanced headphone return, plugged into the camera, so you can monitor the camera and its ability to record/accept your input.

You mean I am hooked up to the camera/cameraman?

Yes, where they go, you go. This is a challenge. The camera is usually up on the cameraperson's shoulder. Their lower profile is that of a normal human being. Your profile, with your mixer slung across your shoulder, or strapped to your chest, is a little wider. Plus all the cables sticking out make you one awkward snagging machine.

Try moving through a classroom full of desks booming at the same time and wrangling that audio snake with no choice but to keep up to the shooter. You are tied together, no? It helps immensely if you know the cameraperson and their style. Just hope you don't get a runner. Or someone who crosses busy streets without thought for their team member. Most audio snakes have quick-release connections which allow you to save yourself from danger.

There is a system that frees each of you from this bondage. It is a pair of wireless transmitters that take the right and left output of your mixer and feeds a pair of receivers on the back of the camera, which in turn are plugged into the camera's inputs. This system is known as "hops," I guess because the audio "hops" from mixer to camera.

That is great! No more cable! What could be better?

Yes, well, it does leave one thing out: the ability to monitor the camera's audio status. If you are separate, then you regularly have to check visually the audio panel on the side of the camera to see if those peak meters are still bouncing. You also have to depend on the cameraperson to listen with an earpiece to hear incoming audio signals.

You now have four more devices to be responsible for. Two transmitters and two receivers. You have to make sure they are powered by either batteries or a DC current feed from the mixer or camera.

Some situations call for the camera mike to be activated on one channel and one channel of wireless to be fed to the second camera input.

Still. You have to be on your toes. Monitor your systems.

So, running around with a cameraman doing sound can be exhilarating. You do have to be your own island of audio goodness.

What does that mean?

You have to be self-sufficient. You have to be a self-contained sound department.

What you need: power. Power. Power.

You have to have fresh cells in your mixer plus a set of backup cells. If your whole rig runs off a central battery pack, you need to carry an extra main battery. This system has a central power distribution unit that feeds your mixer, your wireless receivers, and your "hops" transmitters to send audio to camera.

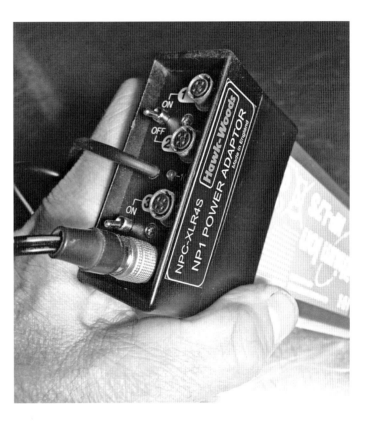

Ecologically it is preferred, because you are changing one big battery again and again instead of filling a landfill with a lot of little dead cells.

It is important to be green.

Save the planet.

It is also important to not have to worry about batteries dying and leaving you unable to do your job.

Save the career.

The great thing about running everything off a single power source is that you only need to worry about the batteries on the talent's wireless transmitter. That's it. You are covered. However, I always make sure that there are fresh cells in all the devices that are being powered by the one block battery, just in case.

The one drawback to the block battery powering all is: If you forget to bring the main battery with you, unless you have stocked all your equipment with fresh cells, you are in trouble.

It happens, leaving the house with your batteries still in the charger. Not if, but when. As a failsafe measure, I put a sign on the door at eye level:

BATTS.

Yes. You can never be too careful.

IFB

On some shows, an additional transmitter is added to your mixer to allow the producers to monitor content.

This is known as an IFB transmitter. It typically is plugged into the tape out socket on your mixer.

It will need batteries.

What does IFB stand for?

Interruptible Foldback, Interruptible Feedback, and Interrupted Foldback, all abbreviated IFB. This stands for, or refers to, a number of systems that allow monitoring of production audio. Without getting into intercoms systems, telephone return feeds, and cueing systems, the most common uses of an IFB is straight monitoring of production audio by the director, producers, script, or clients. (More on IFBs below.)

This additional transmitter, taking a program feed from your mixer, extends your audio, giving all with access to your IFB's frequency the ability to listen to what's going on via a little radio, or a specific monitoring device.

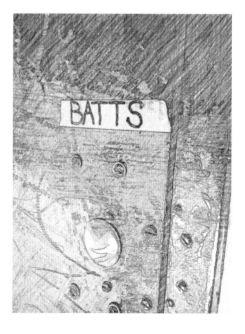

Back to batteries

Are you following one or a number of subjects that are wearing body pack transmitters which you are monitoring for recording? Then you need fresh cells for each of the body pack transmitters. You need to have them on your person/in your sound bag so you can do a quick swap of a weak cell with a fresh one.

Do the body pack transmitters need to be shifted to other subjects? Then you need to carry a kit of supplies to stick the mikes on the talent and protect them from wind.

Are you working at night or indoors in a dark environment? You will need a flashlight with backup cells. Plan your choice of light to be powered by cells that match your most common type. Even if it is the brightest summer day, you might find yourself in a darkened theater or studio, barely able to see a hand in front of your face.

Remember that you are part of a team. At times you might carry additional gear for the camera. A battery. Some media. Perhaps a white card for color balance. A lens cleaning cloth or papers. Be helpful.

MULTI-CAMERA INTERVIEW

Broadcast TV/cable features the correspondent/subject interview more often than the sun rises and sets, so let's take a walk through this genre:

There are two to three cameras.

There is one sound recordist with a mixer.

The mixer just mixes and distributes audio.

Let's go through this:

- Correspondent: Channel 1 (of both cameras)

- Talent: Channel 2 (of both cameras)

There are two booms or lavalieres on both.

Your mixer feeds both cameras, two channels.

The producer wants you to feed the third camera. No problem. Out comes the splitter.

The producer wants a transcription recording of the interview with timecode on one track.

What is a transcription recording?

When a story is being shot, oftentimes interviews are transcribed into printed text so the producers can decide what facts to present without having to watch hours of footage.

Your recording is sent to a transcriber who listens and types it out word for word. The timecode is to allow the producers to return to the footage find the line of text from the transcription.

Sometimes for breaking news stories where time is critical, the mixer will be asked to provide a link to a transcription service. This is basically you calling via a landline phone, a service that will record for transcription in real time the interview you are mixing. Sometimes timecode is involved (see Chapter 11). You will take a small microphone and attach it to the phone earpiece, or route your audio feed through a specialized audio/phone interface box.

For now, we will stick to the actual process of recording sound.

FEATURE FILM OR EPISODIC TV SERIES

The episodic shows we watch usually have a crew of three people:

The mixer

The mixer is the department head. He is the captain of the ship. The team works under his direction. It is usually his gear and his show.

This means he has gotten the assignment by the producers to mix the show/series/film. He chooses who his team is. He makes decisions, with input from his team, on how to record a scene. Usually he chooses the type of microphone to be used, and how a scene will be miked. He instructs the boom operator(s) about frame lines, and helpfully critiques them on their work. Ultimately he is responsible for the location sound on the show. He must not only deal with this responsibility, he must also navigate the often complex political forces and personalities of the production's hierarchy.

He sits (or sometimes stands) behind the mixer recording in multiple ways on multiple devices the sound of the actors' dialog.

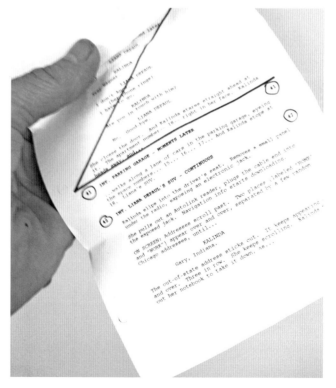

Many mixers us the "sides," which are miniature copies of the script pages to be shot that day to help guide them as they mix scenes.

"Sides"

It is the duty of the utility person to get copies of the sides for his team, and then, if the mixer requests, highlight the different actors' lines in a color unique to them. The lead male's lines are highlighted in blue (e.g. Sam), the lead female's in pink (e.g. Michelle), and so on. Over the course of the shoot, the mixer learns who is what color. The faders on his mixing board has the actors' names to assist the mixer as he does his live dailies mix during the shoot.

The dailies mix is a mono mix of the dialog recorded as it appears on screen. This is often distributed to a separate recording device. It is also what is fed to the IFBs all over the set, to the director, script supervisor, producer, and so on.

Very often, production ends up using this mix for the finished film even though all tracks are being recorded separately on another device.

In addition to the main task of mixing and recording awesome tracks, the mixer must staff his department, represent his department to production, and be a positive person to work for.

The boom

The unique position of the boom operator was outlined in Chapter 6. As you remember, the boom is an essential role that must, with great ability, follow the actors to get their lines clearly and naturally, and stay out of the frame. No reflections, no shadows, or mikes dipping into the shot are the baseline for this job. Another essential duty is to be the eyes and ears of the department. While the mixer is dealing with his high-tech array of gear, and the utility is foot-foaming actors, the boom is watching the rehearsal, and generally keeping an ear to the ground as to what is happening next so the sound department can get a jump on things.

Since the boom is always on set, they must have good relations with the working crews of camera, grip, and electric. They must also communicate with script to update scene numbers and circled takes. The boom is the face of the department.

The utility

This is a rather unsung position. The utility is a busy person.

They ensure that at the beginning of the day:

• power is run to the sound cart

• sides are obtained

• Comteks/IFBs are refreshed with new cells, tested and distributed to those that need them

- a small speaker, if needed, is placed on the video village cart

- slates and Lockit boxes (a small external sync generator) are jammed and distributed to the camera department

- a video feed is run from video village to the mixer's monitors so he can watch exactly what the camera(s) see

- a second boom is built for possible future use

- wireless transmitters are refreshed with new cells, and the corresponding lavalieres are prepped for wiring

- information is gathered as to what the first and second shot of the day will be.

At the end of the day, much of the above is reversed.

- Comteks/IFBs are retrieved, as are the slates.

- All power and video lines retraced back to the sound cart.

- Call sheets for the next day are obtained for all members of the sound team.

IFB devices are necessary tools for the script person and for the director to monitor the actor's lines and performance. You always have a dedicated IFB or Comtek for them. Producers, writers, the main camera operator (so he can move the camera on specific lines), and a host of other people actually have reasons for an IFB. Many people on set want (but often don't need) an IFB and headphones. They are distributed at the beginning of the day either personally or placed on the individual directors' chairs. Sometimes extras are left on the front of the video assist playback cart. You will typically distribute 5–15 IFBs during the day.

Before distribution, they must all be checked to see if they work, and if the batteries are fresh.

Careful note must be taken to count how many go out and if possible, which number went to whom, so at the end of the day, when people vanish like cats, you know who to track down.

People often forget they have them, leave them in odd places, take them home in their bags, put them down somewhere on set . . .

For a piece of equipment worth around $500 to be treated like an empty coffee cup has always been somewhat puzzling to me.

More than once, typically on a commercial, we will be rolling the first take of the day, I will be hard at work at my mixer and someone will appear standing next to me pointing with both hands at their ears. As if I should stop my primary task of mixing, drop everything, and set them up with a Comtek so they can wear it around their neck the rest of the day as set jewelry.

I shouldn't get so frustrated, for we must remember that the IFB/Comtek exchange is one of the few places that the technician and the non-technical meet. Usually with mixed results.

Lectrosonics R1a IFB receiver

Playback operator

The playback operator is responsible for playing music during a scene to motivate the talent as they dance, lip sync, play instruments . . .

Their package usually consist of a laptop, whose output runs to an amplifier, and then that amplifier runs to speakers, either powered or dynamic.

The audio for playback is obtained from production as a file, which is then manipulated by a given program, such as ProTools, for partial or whole playback. It could be a simple song, or an entire orchestra, but either way there are two things you must have to be successful with playback.

The playback cart (left) next to the mixer's cart

- A click track before the first note of the song is played is vital for clean starts from the top of the piece.

- A backup copy of the program material has to be on set. Just in case. At least one backup copy, if not two, of the piece to be played is really essential.

REALITY TELEVISION

This hybrid has the crew size of a feature, but works in a similar way to a documentary, with a game show tacked on.

The **sound supervisor** runs the show. They determine the equipment package to suit the various show, then supervises the installation and testing of the gear. Crews may be selected by the supervisor as is also the case for schedules.

The typical crew is broken down this way:

Sound supervisor

A1   (His right-hand man/woman)

A2   One or more expert technicians who can participate in maintaining the audio infrastructure, wire talent, and fill in for mixers.

Mixers

Any number of documentary style mixers who are paired with cameramen to shoot and mix the show.

The varieties of this type of programming are endless, and other than mentioning the number of jobs available, there is no real fixed model. It is still a medium very much in transition.

HOUSE OF WORSHIP

A less than obvious sound occupation is the mixer of reinforcement audio in a house of worship. The preacher/rabbi/priest/guru and the choir have to be heard by the congregation. This is usually done via speakers in the venue and by feeding other rooms in the facility, and mixed for possible video/radio broadcast video recording and DVD production. Think about it. The house of worship has it all:

- wireless mikes for the pastor
- podium mikes
- pipe organ

- permanent microphone installation for the choir

- miked instruments in the band, with their outputs mixed

- video/web feeds for broadcast/distribution.

That's a lot of audio going on. Now, think about how many churches are in your city . . .

POINTS TO REMEMBER

From a one-man news crew to a big feature film:

- you must be thorough

- you must be prepared for anything

- be professional, but have fun. It's a great job.

14 STEREO:
When Two Become One

"Stereo is merely the attempt to create the illusion of reality through the willing suspension of disbelief."

Richard Heyser

Now that you have mastered the recording of brilliant audio tracks, it's time to open the doors on the world of stereo. The stereo effect is realized by a brilliantly clever series of combinations of microphones, each combination or *array* is chosen for specific applications. Prepare to enter the fascinating world of STEREO.

Let's talk about stereo.

The thing other than mono?

Right. What is stereo?

Wait, what is mono?

One track of recorded sound.

What is stereo?

How many ears do you have?

Well, two of course.

Right. We humans hear, or are used to hearing, from our two ears, which, very much like the way our eyes create depth, create an illusion of directionality and audible perspective. Add delay of arrival triangulation stereo.

What?

Listen, sound arrives at your ears at slightly different times.

That's how we perceive an audible perspective, just as our two eyes can see differences in distance and "see" three dimensions. We hear and judge the arrival times of the sound patterns in our world, and "hear" stereo.

A stereo recording is an electromechanical attempt to replicate how we hear with two ears that are separated by our big head.

Using two or more independent audio channels played or reproduced through two or more loudspeakers or a pair of stereo headphones, we create the impression of sound heard from a wide sound field just like the way we hear with our ears.

You would hear different things on different sides of the listening spectrum, your right/left headphones, or speaker array, but they are all mono tracks mixed in a particular way to a specific side.

In fact, one type of stereo array, *not a stereo microphone—* for stereo always requires at least two microphones (or two transducers in a single unit)—is a replica of a human head in size, density, and shape, with two microphones (transducers) in the position of each eardrum.

The Germans have a name for it: *kunstkopf*. This oddly translates into: *art head*.

From the Neumann catalog:

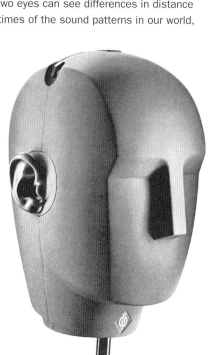

The Neumann KU 100

"The KU 100 dummy head is a binaural stereo microphone. It resembles the human head and has two microphone capsules built into the ears. When listening through high-quality headphones it gives the illusion of being right at the scene of the acoustic events."

This is an example of what engineers call "true stereo."

It is a recording simultaneously using two microphones placed in a specific array pointed at the sound source. These mikes produce two similar tracks but each subtly different due to time or arrival difference of the sound waves, and slight differences in the arriving sound pressure as well. Just like we hear. When you play back the separate recordings through a right/left set of speakers or headphones, you, the listener, hear the subtle differences in timing and sound level and triangulate mentally the positions in space of whatever is producing noise.

Basically, since the moment Alan Blumlien became dissatisfied during a screening of one of the first "talkies," stereo sound has attempted to replicate the way humans hear sound for the screen. Blumlien was watching a film in a cinema with his wife, in which the characters moved from one side of the screen to the other. He began to think about realism with sound, and afterward remarked to his wife that he could create the illusion of movement across the screen through sound. He proposed the idea to his boss at EMI, and went to work. His patent submission was, "Improvements in and relating to Sound Transmission, Sound-recording, and Sound-reproducing systems." This patent was dated December 13, 1931. Many of his concepts are still used today. It is notable that this research was first focused on motion picture sound recording and playback.

The microphone technique Blumlein employed was naturally named and is still known as a "Blumlein Pair" or as they say in England: a "Blumlein Array." In more familiar terms it is known as an *X/Y pair*.

The Blumlein Pair is basically two crossed figure-of-eight microphones. This array produces an exceptionally realistic stereo image.

In actual film production work, the use of figure of eight microphones can be limited by practical issues. The figure of eight, either condenser or ribbon is very sensitive to wind when used outdoors.

However, other popular stereo arrays using Blumlein's X/Y coincident technique are a hypercardioid or cardioid pair.

Alan Dower Blumlien, 1903–1942

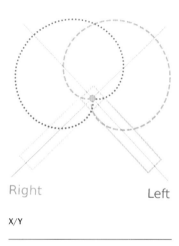

Right   Left

X/Y

Two Schoeps Mk 41 hypercardioid capsules in an X/Y Pair

Two hypercardioid or supercardioid mikes which produce a great stereo image across a narrower recording angle.

MS MID-SIDE STEREO PAIR

Another contribution Alan Blumlein made to audio is the M/S Mid-Side pair. Patented in 1931, like his X/Y pair, this employs two microphones; however, they are very different microphones rather than the identical pair used in the X/Y array.

M/S uses a cardioid and a figure of eight.

The mid-microphone is typically a cardioid (one can, for alternative imaging, use an omni).

The side microphone must be a figure of eight situated so its active pickup pattern is 90 degrees off-axis from the sound source.

Both capsules must be placed as closely together as possible, usually one above the other.

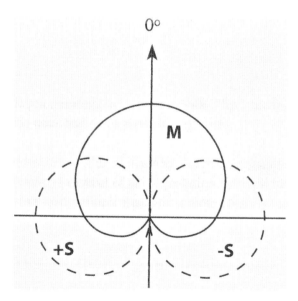

Unlike X/Y, which once recorded does not allow much to work with in altering your stereo image in post, M/S allows much greater flexibility.

Its stereo image is directly dependent on the amount of input coming from the side channels, so raising or lowering the ratio of mid to side will create a broader or narrower stereo field.

This means you can alter the sound of your recording in post, something you cannot do with X/Y.

How does it work?

It's not that difficult. Look at the cardioid mike pointed at the sound source as the center channel picking up a clear recording of the sound source.

Look at the figure of eight as the side channel, picking up ambience and reverberant sound from the sides of the sound source.

In post, you can process the side channel in various ways to create two tracks. We will not go into post processing in this chapter, just understand that a few easy steps can give you a versatile and malleable stereo recording.

For broadcast situations where mono is required, M/S is quite adaptable. All you need do is switch the mix to mono which nullifies the figure of eight signal (through phase cancelation), leaving only the cardioid to carry the load. The best of both worlds.

OTHER STEREO SYSTEMS

ORTF
This was invented in the early 1960s by *Office de Radiodiffusion Télévision Française* (ORTF) at Radio France.

The two cardioid or hypercardioid microphones are set at a 110-degree angle spaced 17 cm apart from each other.

The result is a realistic stereo field that has manageable compatibility with mono. (Television in 1960 was broadcast in mono.)

Audio engineers could record stereo for disk release and use the identical recording for broadcast.

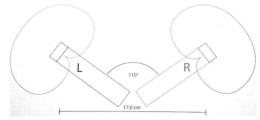

N.O.S. Array
Netherlands Radio engineers developed the N.O.S. array (Nederlandsche Omroep Stichting), which also employs two cardioid microphones like the ORTF system. The major differences are the angle of the microphones (90 degrees) and the greater distance between them than the ORTF. In the NOS array they are 30 cm apart (versus ORTF, 17 cm apart as mentioned above). Less angle, more distance.

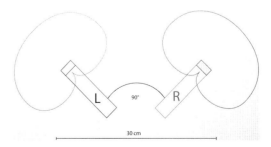

Differences are noticeable in the lower frequencies. The greater distance also makes this system less mono-compatible.

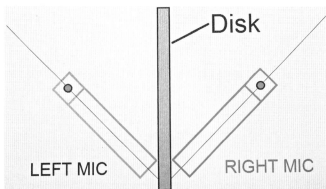

The OSS/Jecklin Disk

Jurg Jecklin of Swiss Broadcasting developed a system called OSS (Optimal Stereo System). This array features a 30 cm (1 ft) disk about 2 cm (3/4") thick which has a layer of sound-absorbing material covering its surface. Microphones are placed above the surface of the disk, just in the center, 16.5 cm (6 1/2") apart from each other and pointing 20 degrees outside. This acoustically opaque baffle separates the microphones as an approximation of the adult human head.

The Decca Tree

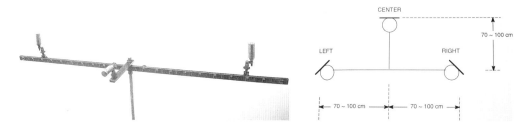

This is a system of three omnidirectional microphones placed at the end of a large T-shaped mount. The left and right microphones are assigned fully to the left and right channels, respectively, while the center microphone is equally to both but at a reduced level to avoid an over-prominence of the sound from the center. Omnidirectional microphones always tend to become more directional at higher frequencies, so the principal axes of the microphones are aimed inward and downward, toward the sound source.

The Decca Tree was developed by Roy Wallace for English Decca Records in 1954. Its first trials were with an orchestra, positioned a few feet behind and about ten feet above the conductor. Arthur Haddy, Wallace's partner, remarked that the rig looked "like a bloody Christmas tree." The name stuck.

The Decca Tree configuration proved so versatile because of its ability to produce a pleasing and stable stereo image that it has spawned a number of variations.

OCT/OCT2 (Optimized Cardioid Triangle)

Dr. Gunther Theile has developed a number of systems for stereo and surround sound which are notable for their simplicity and great sound quality.

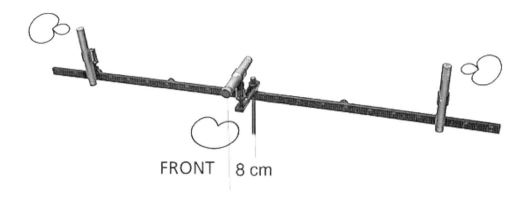

FRONT | 8 cm

OCT

The first is a variation on the Decca Tree employing microphones in a right/left/center configuration. This array is notable for the types of microphones and distances between them. In the classic Decca Tree the microphones are omnidirectional. In the OCT, Dr. Theile uses three directional microphones. The center channel sound is picked up by a cardioid. It is mounted on the same plain as the right and left microphones. The right and left channels are supercardioids mounted 90 degrees off axis of the center cardioid. The right and left microphones, due to their high directionality, pick up front-incident sound at much lower levels.

The strength of this area is its good separation between the center/left and center/right sectors. The listening area is enlarged and multiple phantom images are avoided, which would degrade the naturalness of the sound.

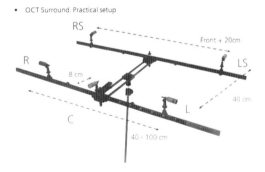

• OCT Surround. Practical setup

OCT 2

This variation is closer to the classic Decca Tree in that the center cardioid is mounted on a bar shifting it forward (40 cm instead of only 8 cm) of the right and left channel supercardioids, giving more of a spacious "Decca Tree" type of sound.

OCT Surround

This is an expanded OCT system using a second mounting bar behind the initial array. The bar, which is 20 cm wider than the front, has two cardioids faced rearward from the sound source. Lateral reflections, which are important for the preemption of the room, are reproduced accurately, producing a convincing spatial perspective.

Hamasaki Square

This system, created by Kimio Hamasaki for NHK, the Japanese National Broadcasting Network, consists of a cruciform mount on which, at the end of each bar, is mounted a figure-eight microphone routed to the left, right, and left surround and right surround channels. It is an ideal system for picking up the diffuse sound field in a reverberant environment.

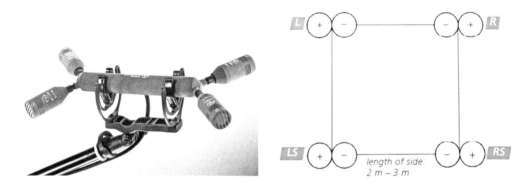

length of side:
2 m – 3 m

The recent popularity and availability of multi-track recorders has allowed access to a field that before was reserved for specialized studio or concert hall sessions. Now, even the most modest handheld SD card recorder has a stereo pair right up on top of the unit. These multi-track recorders are a valuable teaching tool for listening and learning about stereo imaging. Some units allow you to easily switch from X/Y to ORTF and back. This feature, as well as their miniature size, makes them worth their modest price.

A great exercise would be to experiment with such a unit recording active sound fields both inside and outside.

Slate the head of each take with the configuration you are using (X/Y or ORTF) and record a minute of sound.

Later, in familiar monitoring surroundings, play back the minute of audio over speakers. Then listen to the same tracks over headphones. Note the differences not only between the microphone configurations but also between the listening environments. It will be a great lesson not only in stereo, but also in the art of listening.

The art of stereo is truly based on listening. The above configurations are starting points for any given recording situation. Altering the angle or distance of a mike a few centimeters in a given space is really your best option, and it is up to your judgement to compensate for the various recording situations in which you may find yourself.

POINTS TO REMEMBER

It is important to listen to any microphone you are using. With stereo it takes another level of perception to understand fully the aural imaging that you hear. It takes practice to know what you are hearing when working with stereo. Start practicing!

APPENDIX

FILMS TO HEAR

Apocalypse Now—Francis Ford Coppola

Berberian Sound Studio—Peter Strickland

Birdman—Alejandro Gonzalez Inarritu

Blow Out—Brian De Palma

Citizen Kane—Orson Welles

Cléo from 5 to 7—Agnès Varda

Dr. Strangelove—Stanley Kubrick

Elevator To the Gallows—Louis Malle

In The Mood For Love—Wong Kar-Wai

La Grand Belizza—Paolo Sorrentino

Pickpocket—Robert Bresson

Red—Krzysztof Kieslowski

Raging Bull—Martin Scorsese

Rumble Fish—Francis Ford Coppola

Saving Private Ryan—Stephen Spielberg

The Birds—Alfred Hitchcock

The Conversation—Francis Ford Coppola

The Hurt Locker—Kathryn Bigelow

The Silence of the Lambs—Jonathan Demme

The Stranger—Orson Welles

THX 1138—Paul Lucas

GLOSSARY

ADR (Automatic Dialog Replacement)—This is when an actor watches a scene shot previously with problematic dialog, performs their lines (and is re-recorded) in a studio situation.

ambience—The sound present in a given location other than the spoken dialog/music or effects sound from on-camera action. Ambience includes any present environmental noise in the location such as wind, birds, air-conditioning, etc. (*see* room tone).

amplifier/amp—An electronic circuit that increases the signal level by increasing the voltage or current. Examples: power amplifiers boost signals to drive loudspeakers. Pre-amplifiers boost small amounts of voltage from microphones.

amplitude—A term describing signal level measured in decibels.

attenuator—A device to reduce signal level.

audio frequency—A range describing a human's perception of sound. Commonly understood at 25 Hz to 20 Hz.

bass—The low-frequency portion of the audio spectrum.

Bell, Alexander Graham—Scientist, engineer, inventor, and innovator generally credited with the invention of the first practical telephone. Bell, a master ventriloquist, taught his Skye Terrier to speak a crude phrase, did years of research in hearing and visual speech, and worked with the deaf. This work led to his invention of the telephone. The Bell Telephone Company, in the early twentieth century, named the measurement sound pressure after Bell: dB = deci(ten)Bel (A.G. Bell) = **deciBel**.

bidirectional—The description of the pickup (polar) pattern of a figure-eight microphone consisting of two circular lobes of equal sensitivity and opposite polarity from the front and back of a ribbon diaphragm.

Blumlein, Alan Dower (1903–1942)—Inventor of fundamental patents involving audio, stereo phonographs, television and radar. Inventor of the Blumlien Pair, which consists of a pair of crossed bidirectional microphones that are arranged at a 90-degree angle to each other so that the principal axis of each microphone is precisely aligned with the null axis of the other.

cardioid—A microphone pattern shaped like a heart, sensitive to the front of the microphone and insensitive to the rear.

channel—A single signal path of audio from microphone through amplifiers and signal processors to reproduction amplifiers and loudspeakers. Mono: single channel. Stereo: two channel.

cochlea—The inner ear. This is the nautilus shaped organ filled with nerve hairs and fluid that is the primary organ of hearing.

coloration—The deterioration of an audio wave form from its purest form to dilution with various forms of reverberation.

condenser microphone—A powered microphone consisting of a diaphragm and a capacitor plate. Incoming sound waves change the spacing between the two membranes. The difference in distance between the plates is transfused to a usable signal voltage.

decibel (dB)—The standard measurement of sound pressure. A combination of deci (ten) and the first three letters of the last name of Alexander Graham Bell.

diaphragm—The term for a moveable membrane, either in a microphone, a speaker or the human ear which responds to, or transfers energy of sound waves.

distortion—(In sound) The warping or alteration of the original shape of a waveform.

dynamic microphone—A microphone that generates electrical current as sound waves move its diaphragm attached to an electrical conductor in a magnetic field.

dynamic range—The spectrum of usable sound between soft and loud. All audio systems are limited at the low end by a noise floor of the

circuits and at the high end by distortion or dropout. Everything in between is the dynamic range of the system.

ear drum—Also known as tympanic membrane, this is a thin cone-shaped membrane that separates the external ear from the middle ear. it transmits sound from the air to the ossicles inside the middle ear, and then to the oval window in the cochlea. Like a microphone, it is a transducer that converts and amplifies vibration in air to, in our case, vibration in fluid.

electret microphone—A microphone similar to the capacitor with its element permanently charged so no polarizing voltage is necessary. Electrets' charged elements are used to good advantage in miniaturized form in the construction of small lavalieres.

Eustachi, Bartolomeo (1514–1574)—One of the founders of the science of human anatomy. He extended the knowledge of the internal ear by rediscovering and describing correctly the tube that bears his name.

Eustachian tube—The tube that equalizes pressure in both sides of the eardrum running between the middle ear and the pharynx.

Foley—The artificial recreation and recording in post-production of sounds for actions seen on-screen/in picture.

frequency—The number of waves in any given frequency over one second. Measured in hertz (Hz).

hertz—The unit of frequency describing the number of cycles per second of a vibrating audio signal. Abbreviated Hz. Named for Heinrich Hertz.

howl around—An alternate term for acoustic feedback.

hypercardioid—Part of a family of microphones created by the combination of the omni and the figure of eight, the hypercardioid sounds similar to the cardioid but with increased sensitivity in front, a narrower pickup pattern and some appreciable sensitivity in the rearward direction.

IFB—Interruptible holdback, interruptible feedback, abbreviated IFB is basically, in TV and filmmaking, a feed of program audio for the monitoring of various departments on set. The director, script, and producers can all listen in on the dialog being recorded.

impedance—The opposition or resistance of the flow of electrical energy, measured in ohms.

inner ear—The cochlea or sound analyzing section of the organ of hearing.

interference—The combining of two or more signals resulting in a negative interaction, which causes the degradation of the quality of one or more of the signals.

Kudelski, Stefan—The engineer responsible for the modern magnetic/analog miniature tape recorder, the Nagra (see Nagra).

magnetic—A property of certain materials to attract one another. This may be introduced by the influence of an electrical current.

microphone—A device that transduces/ transforms vibrations in the atmosphere into corresponding electrical signals.

microphone polar pattern—The directional character of a microphone graphically mapped indicating the microphone's directivity and how it responds to sound energy shown in a diagram of the horizontal plane through 360 degrees.

middle ear—The part of the ear between the drum and the cochlea.

MIDI (Musical Instrument Device Interface)—A popular interconnection bus for synthesizers, drum machines, and other musical devices capable of electronic control.

mike—Slang for microphone.

mix—(n.) The completed combination of various tracks or inputs to create a single soundtrack.

(v.) The act of combining various inputs/tracks to create a single soundtrack.

monophonic—A single channel recording arranged for two-eared listening.

monoral/mono—A single channel recording or reproducing system.

Nagra—A company that produced a series of high-quality reel-to-reel analog tape machines that revolutionized sound for film due to their compactness, quality, and versatility (see Kudelski, Stefan).

noise floor—The inherent noise present in a circuit or system that establishes the lowest audible usable signal level.

Omnidirectional—(In audio) The description of a polar pattern which responds to sound energy nearly equally from all directions.

Optical sound—The process used to record sound photographically via a photoelectric cell responding to audio energy, usually using a variable area to present the audio waveform.

ORTF system—*Office de Radiodiffusion Télévision Française* (Radio France) devised a stereo miking system using two cardioid microphones spread to a 110-degree angle.

ossicles—The three bones of the middle ear (incus, malls, stapes) that provide a physical linkage between the eardrum and the oval window of the cochlea.

peak meter or PPM (peak program meter)—A meter indicating levels that responds quickly to incoming signals.

phase—The time relationship between two signals.

pinna—The outer ear. Directional cues can be aided by the folds of the pinna.

pitch—The subjective perception of frequency.

polar pattern—See microphone polar pattern.

potentiometer—A voltage divider used for measuring electric potential (voltage). Commonly

217

used as volume controls on electronic equipment, or if operated by a mechanism used as position transducers, like a joystick.

pressure gradient—A term usually given to a transducer (figure-of-eight microphone) that responds to differences of pressure on both sides of the ribbon of the microphone.

pressure microphone—A microphone that responds only to the changes of air pressure at the diaphragm. A pure pressure microphone has an omnidirectional polar response.

rarefaction—The part of the sound wave where the air particles are spread apart (uncompressed).

reverberation—A temporary extension of an acoustical event (sound wave) created by multiple sound reflections.

ribbon microphone—A microphone that converts acoustical energy into electrical energy through the use of a fine metallic ribbon placed in a magnetic field. The ribbon is moved by sound waves, which creates a signal voltage between its ends. Operating on the differences of pressure between its two sides, this results in a figure-of-eight polar pattern.

room tone—The recording of the interior of a set or room in which production has just taken place. This ambience, a very valuable tool, usually of 20–30 seconds' duration, is used by editorial to bridge gaps in dialog, cover noises, and generally help in cutting dialog.

shotgun microphone—A microphone using an interference tube (differential phase) to achieve a highly directional pattern. Generally a hypercardioid or supercardioid microphone.

signal-to-noise ratio—A term describing the level of a desired signal to the level of background noise that a given device or signal path produces. High signal-to-noise (S/N) is good. A low signal-to-noise ratio is not desirable.

sound design—The process of recording, selecting, and mixing sounds together to produce a desired artistic effect for a production. Typically referred to as a post-production action.

sound mixer—(n.) An electronic device used to process and distribute acoustic information. (v.) A person who is a production sound recordist.

sound pressure—The main element used to measure the amplitude of sound waves. The level of sound pressure is usually measured in decibels.

soundtrack—(1) Used to describe the physical placement of audio information on a tape or film print. (2) The audio component of a visual presentation.

standing waves—A room resonance, usually in the low frequency, that results in interaction between waves leading to a mutual reflection at similar frequencies. The uneven distribution of sound energy creates an unpleasantly flat resonance.

stereo—The Greek word meaning solid, indicating depth, breadth, and height. The aural ability to perceive depth due to time-of-arrival differences in sound waves heard by a listener.

stereophonic—A term describing any number of audio systems that can represent sound spatially.

sticks—Slang term for slate or clapperboard.

supercardioid—A microphone polar pattern which is an exaggerated hypercardioid (see hypercardioid).

sync sound—Sound record with picture. Either single system, where the audio information is recorded with the visual medium, or double system, where the audio information is recorded on a separate device and *synced* later.

tone—(1) A sound that is identifiable by its constant pitch. (2) A generated reference signal from a mixer to a record device to set levels and inform post.

track—The physical area on a medium (tape, disk, film, program) within which audio data is recorded or printed.

transducer—A device that transforms one form of energy into another. In audio usage, the most common forms are the microphone and the loudspeaker.

VU meter—A meter intended to read the signal level in audio equipment in a standardized fashion. Based on 1940s electromechanical technology, it is considered obsolete as a standalone device for level measurement.

wavelength—The measurement between consecutive corresponding points of the same phase. The wavelengths of sound frequencies audible to the human ear (20 Hz–20 kHz).

wild sound—Sound recorded during a production that is gathered independent of a camera/picture for use later in editing/sound design.

X/Y stereo pair—Also known as the Blumlein Pair (see Blumlein, Alan Dower). Originally created using microspaced figure-of-eight microphones placed at a 90-degree angle to one another. Applications can include cardioid and hypercardioid microphones, still at 90 degrees. The sonic image produced by X/Y is considered by many to create a realistic stereo sound image.

INDEX